GENERAL HANDBOOK OF ON-LINE PROCESS ANALYSERS

ELLIS HORWOOD SERIES IN ANALYTICAL CHEMISTRY

EDITORS: Dr. R. A. Chalmers and Dr. Mary Masson, University of Aberdeen

"I recommend that this Series be used as reference material. Its Authors are among the most respected in Europe". *J. Chemical Ed.*, New York.

Application of Ion-selective Membrane Electrodes in Organic Analysis
F. BAIULESCU and V. V. COSOFRET, Polytechnic Institute, Bucharest
Industrial Methods of Microanalysis
S. BANCE, May and Baker Research Laboratories, Dagenham
Ion-selective Electrodes in Life Sciences
D. B. KELL, University College of Wales, Aberystwyth
Inorganic Reaction Chemistry
Volume 1: Systematic Chemical Separation
D. T. BURNS, Queen's University, Belfast, A. G. CATCHPOLE, Kingston Polytechnic,
A. TOWNSHEND, University of Birmingham
Volume 2: Reactions of the Elements and their Compounds
D. T. BURNS, Queen's University, Belfast, A. TOWNSHEND, University of Hull, and
A. H. CARTER, North Staffordshire Polytechnic
Handbook of Process Stream Analysis
K. J. CLEVETT, Crest Engineering (U.K.) Inc.
Automatic Methods in Chemical Analysis
J. K. FOREMAN and P. B. STOCKWELL, Laboratory of the Government Chemist, London
Fundamentals of Electrochemical Analysis
Z. GALUS, Warsaw University
Laboratory Handbook of Thin Layer and Paper Chromatography
J. GASPERIC, Charles University, Hradec Kralove
J. CHURACEK, University of Chemical Technology, Pardubice
Handbook of Analytical Control of Iron and Steel Production
T. S. HARRISON, Group Chemical Laboratories, British Steel Corporation
Handbook of Organic Reagents in Inorganic Analysis
Z. HOLZBECHER et al., Institute of Chemical Technology, Prague
Analytical Applications of Complex Equilibria
J. INCZEDY, University of Chemical Engineering, Veszprém
Particle Size Analysis
Z. K. JELINEK, Organic Synthesis Research Institute, Pardubice
Operational Amplifiers in Chemical Instrumentation
R. KALVODA, J. Heyrovský Institute of Physical Chemistry and Electrochemistry, Prague
Atlas of Metal-ligand Equilibria in Aqueous Solution
J. KRAGTEN, University of Amsterdam
Gradient Liquid Chromatography
C. LITEANU and S. GOCAN, University of Cluj
Titrimetric Analysis
C. LITEANU and E. HOPIRTEAN, University of Cluj
Statistical Methods in Trace Analysis
C. LITEANU and I. RICA, University of Cluj
Spectrophotometric Determination of Element
Z. MARCZENKO, Warsaw Technical University
Separation and Enrichment Methods of Trace Analysis
J. MINCZEWSKI et al., Institute of Nuclear Research, Warsaw
Handbook of Analysis of Organic Solvents
V. SEDIVEC and J. FLEK, Institute of Hygiene and Epidemiology, Prague
Methods of Catalytic Analysis
G. SVEHLA, Queen's University of Belfast, H. THOMPSON, University of New York
Handbook of Analysis of Synthetic Polymers and Plastics
J. URBANSKI et al., Warsaw Technical University
Analysis with Ion-selective Electrodes
J. VESELY and D. WEISS, Geological Survey, Prague
K. STULIK, Charles University, Prague
Electrochemical Stripping Analysis
F. VYDRA, J. Heyrovský Institute of Physical Chemistry and Electrochemistry, Prague
K. STULIK, Charles University, Prague
B. JULAKOVA, The State Institute for Control of Drugs, Prague

GENERAL HANDBOOK OF ON-LINE PROCESS ANALYSERS

D. J. HUSKINS

Senior Instrument Engineer
W. G. Engineering Design Limited
Aberdeen

ELLIS HORWOOD LIMITED
Publishers · Chichester

Halsted Press· a division of
JOHN WILEY & SONS
New York · Brisbane · Chichester · Toronto

First published in 1981 by

ELLIS HORWOOD LIMITED

Market Cross House, Cooper Street, Chichester, West Sussex, PO19 1EB, England

The publisher's colophon is reproduced from James Gillison's drawing of the ancient Market Cross, Chichester.

Distributors:

Australia, New Zealand, South-east Asia:
Jacaranda-Wiley Ltd., Jacaranda Press,
JOHN WILEY & SONS INC.,
G.P.O. Box 859, Brisbane, Queensland 40001, Australia

Canada:
JOHN WILEY & SONS CANADA LIMITED
22 Worcester Road, Rexdale, Ontario, Canada.

Europe, Africa:
JOHN WILEY & SONS LIMITED
Baffins Lane, Chichester, West Sussex, England.

North and South America and the rest of the world:
Halsted Press: a division of
JOHN WILEY & SONS
605 Third Avenue, New York, N.Y. 10016, U.S.A.

© 1981 D.J. Huskins/Ellis Horwood Ltd.

British Library Cataloguing in Publication Data
Huskins, D.J.
General handbook of on-line process analysers. —
(Ellis Horwood series in analytical chemistry)
1. Chemistry, Analytic — Laboratory manuals
I. Title
543'.0028   QD76

Library of Congress Card No. 81-13485 AACR2

ISBN 0-85312-329-2 (Ellis Horwood Ltd., Publishers)
ISBN 0-470-27292-9 (Halsted Press)

Typeset in Press Roman by Ellis Horwood Ltd.
Printed in Great Britain by R. J. Acford, Chichester.

Table of Contents

Chapter 15 – Analytical Method Selection Guide

Chapter 16 – The Economic Justification for Installing an Analyser 234

Foreword

It is generally believed to be true that the laboratory analyser of today becomes the process analyser of tomorrow, but sometimes it is a long day before the transformation is affected. The history of the glass electrode as a laboratory device for the measurement of pH is an example of a slow metamorphosis starting from the early discovery of the properties of a glass membrane by Cremer about 1906 and reaching a satisfactory industrial stage of development about 1940. The history of gas chromatography showed a faster pace following the pioneer development for laboratory use which took place in 1950 and the subsequent development of the on-line analyser less than a decade later.

There are several reasons for the long delay which frequently occurs before an industrial analyser emerges from a laboratory technique. In the first place, the physicist or chemist who developed the technique is generally neither interested nor capable of extending the development of the industrial application. This step has to be made by technologists experienced in the difficulties and requirements of continuous process measurements and such people tend to be rare in comparison with the armies of laboratory physicists and chemists, intent on developing new types of laboratory techniques and methods of analysis. Perhaps this explains the comment by the author, D. J. Huskins, that there are many good specialist books on the various analytical techniques and methods but only a few which have been written to describe process analysers. Therefore, this series of books makes a most welcome contribution towards redressing the balance.

A second reason for the delay in the emergence of process analysers from the laboratory stage is linked with the difficulty of obtaining a continuous sample for analysis which is not only representative of the main stream but has to be delivered to the instrument with minimum delay, without significant change in composition or physical state and at a level of cleanliness with which the analyser can cope. The importance of sampling techniques and systems, including sample handling and conditioning for process analysers, could justify the devotion of a complete book to the subject.

The third and most important reason for the cautious and often delayed introduction of laboratory techniques for process analysis is linked with the heavy burden of the dependability that rests on the process instrument. It is not the capital cost of the instrument which is important — this is frequently far outweighed by the dependence on the instrument for process efficiency and throughput, particularly if there is an automatic control function linked to the measurement. An accurate reliable analytical instrument on a high-throughput process can be a priceless asset. An inaccurate unreliable instrument can be a menace of the worst type which could lead to losses through lower efficiency or, worse still, to a plant shut-down. In certain cases it could lead to plant damage or even loss of life. This burden of responsibility rarely rests on a laboratory instrument and hence the need for caution and conservatism in the introduction of new process analysers.

From this it is clear that there should be wide understanding of the operation characteristics and limitations of analytical instruments used for industrial process control. The Author, who has had a long practical experience with process analysers in the field and has shown great courage, capability and assiduity in covering such a wide field of measurement so comprehensively deserves much commendation for making his experience and knowledge available on a subject of great importance to industrial safety and efficiency.

R. S. MEDLOCK

Preface

This book was originally intended to form part of a single-volume work on process analysers, but the proposed work became so long that it was divided into five separate books, of which this is the first. The others are Quality Measuring Instruments (Book 2), Optical Methods (Book 3), Electrical and Magnetic Methods (Book 4) and Gas Chromatographic Methods (Book 5). Books 2–5 contain descriptions of the various methods, instruments available, and some applications. Other applications are given in this volume, which also gives the general abbreviations, symbols and units used in all the books. Other symbols etc. may also appear if they are widely used in a particular industry, but international agreement on units and symbols, and consistent use of them, are essential to avoiding *mis*understanding (which can be more dangerous than *not* understanding). This first volume does not contain detailed descriptions of analysers, but is intended to give the reader a better understanding of the proper use of analysers. The list of contents should give the reader a good idea of what should be considered when an analyser system is to be installed. The chapters on economic justification and on the selection of analytical methods are located at the end of the book, but should be amongst the first chapters examined when the installation of an analyser system is first considered.

Most instrument engineers enjoy their work because of the great variety of problems associated with instrumentation. Analyser system engineering is an extension of instrumentation that I have found particularly interesting. I therefore hope that the reader will benefit from reading the series and will find analyser system engineering as interesting as I have.

I would like to thank R. S. Medlock of Kent Instruments Ltd., for his invaluable comments and for writing the Foreword, and also to acknowledge the help of the various manufacturers who have supplied me with considerable information about their products and kindly given permission to reproduce illustrations (without which I could not have written the book). Finally I would like to acknowledge the encouragement and hard work of N. Goodman of Adam Hilger Ltd., and R. A. Chalmers and Mary R. Masson of Aberdeen University, and of the staff of Ellis Horwood Ltd. who have patiently produced the series.

D. J. HUSKINS

Graphic Symbols and Abbreviations

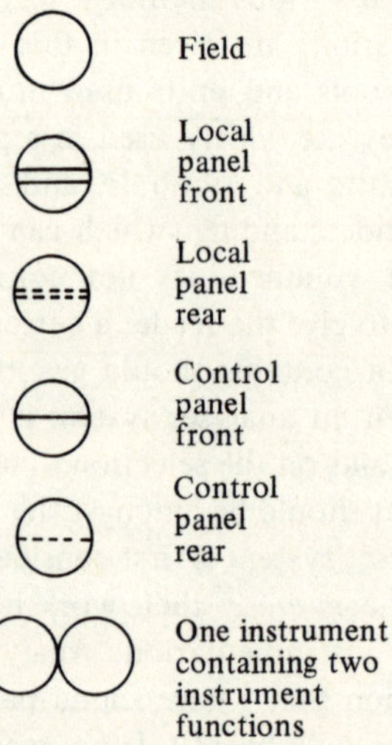

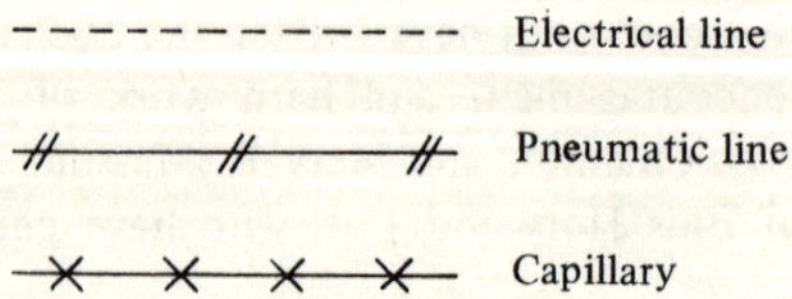

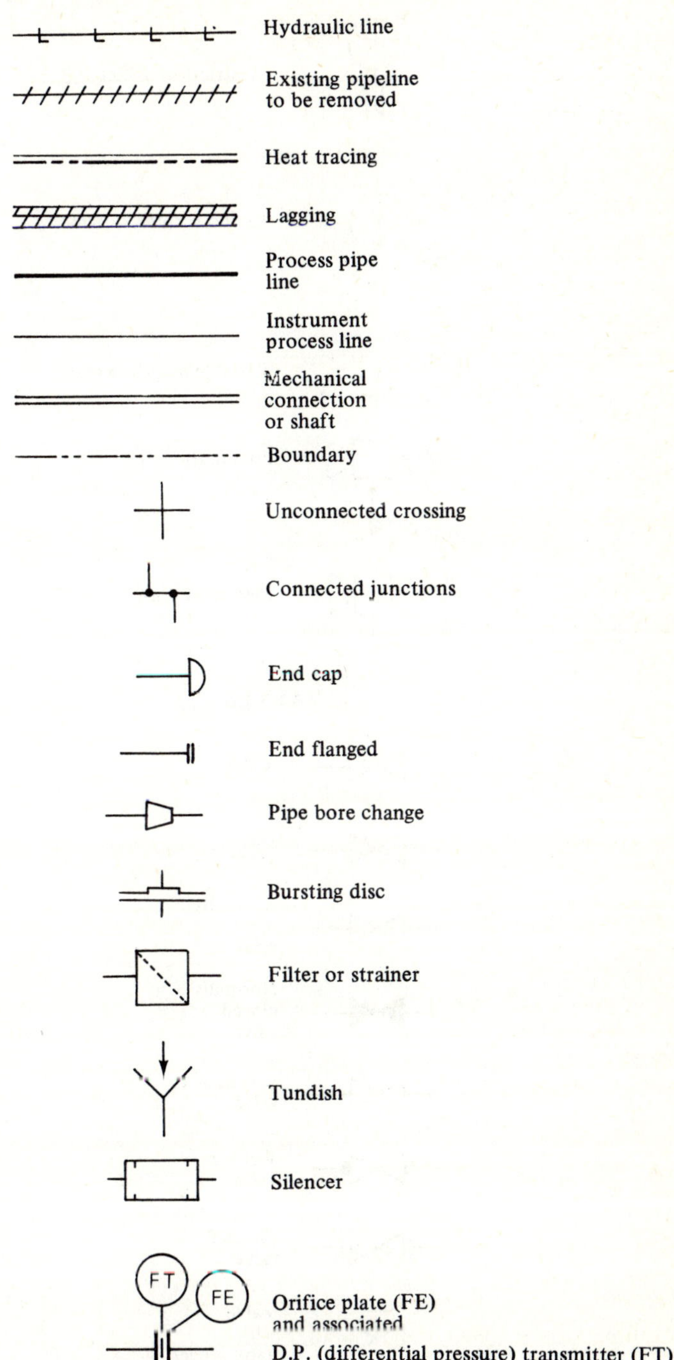

Hydraulic line
Existing pipeline to be removed
Heat tracing
Lagging
Process pipe line
Instrument process line
Mechanical connection or shaft
Boundary
Unconnected crossing
Connected junctions
End cap
End flanged
Pipe bore change
Bursting disc
Filter or strainer
Tundish
Silencer
F T
FE
Orifice plate (FE) and associated
D.P. (differential pressure) transmitter (FT)

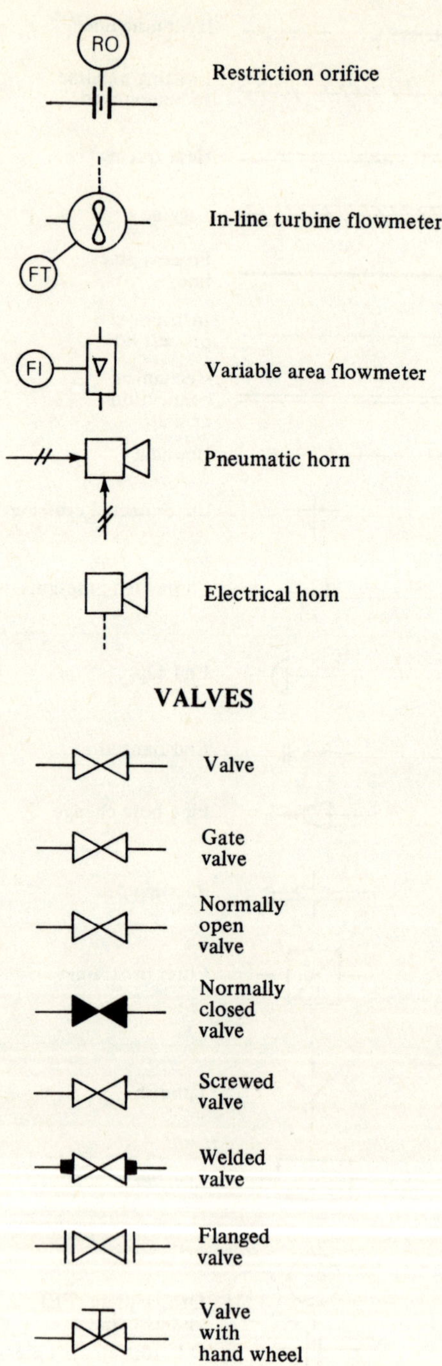

Restriction orifice

In-line turbine flowmeter

Variable area flowmeter

Pneumatic horn

Electrical horn

VALVES

Valve

Gate
valve

Normally
open
valve

Normally
closed
valve

Screwed
valve

Welded
valve

Flanged
valve

Valve
with
hand wheel

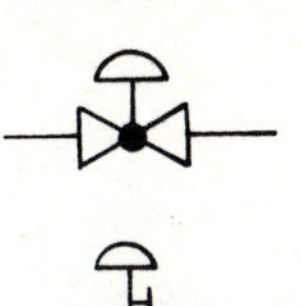

Diaphragm
operated
globe
valve

Diaphragm
operated valve
with hand wheel

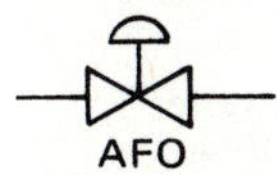

Air
failure
open

Air
failure
closed

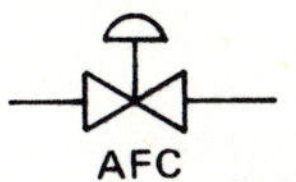

Air
failure
lock

Control valve
in process flow
diagram

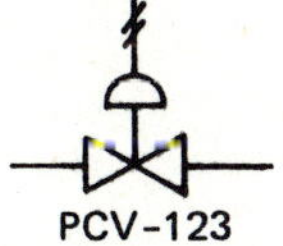

Self operated
valve.
Regulator or
reducer
Direction of
flow ⟶

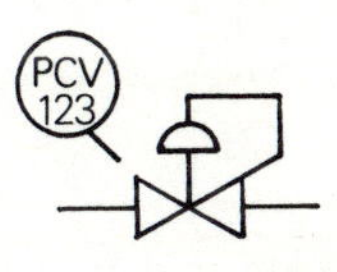

Self operated
valve.
Back pressure
regulator
Direction of
flow ⟶

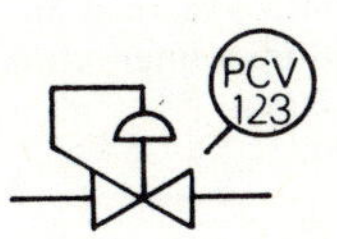

Diaphragm
operated
angle
valve

Diaphragm
operated valve
with positioner
showing air
supply

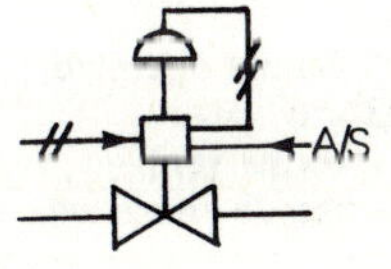

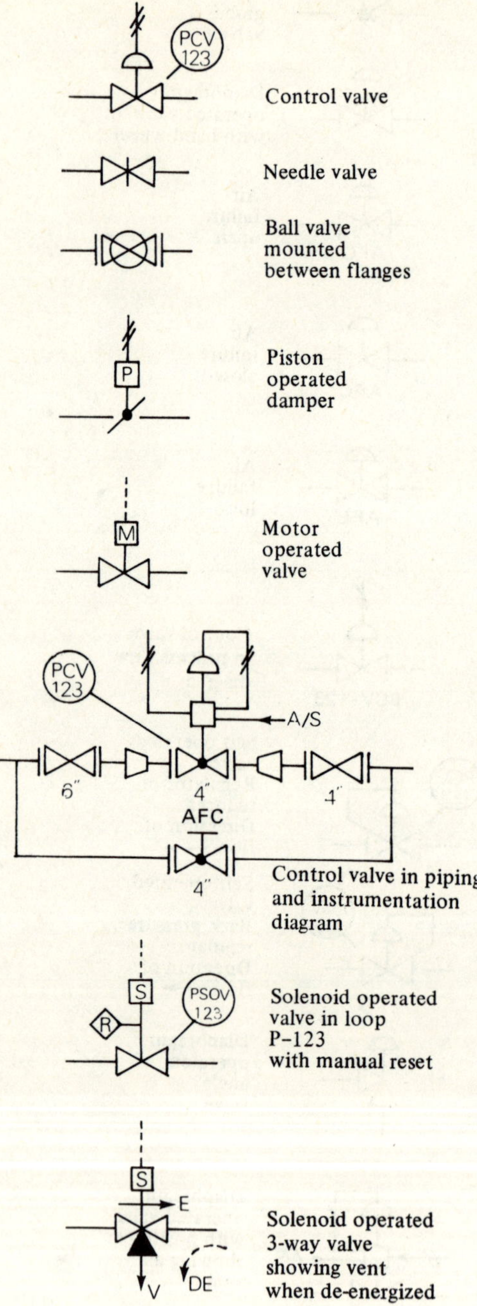

Control valve

Needle valve

Ball valve
mounted
between flanges

Piston
operated
damper

Motor
operated
valve

Control valve in piping
and instrumentation
diagram

Solenoid operated
valve in loop
P-123
with manual reset

Solenoid operated
3-way valve
showing vent
when de-energized

Check valve

14 bar ga

Pressure relief
valve set at
14 bar ga

Vacuum relief
valve

Butterfly valve
between flanges

Cylinder gas
pressure regulator
fitted to a
cylinder

Example of use of abbreviations (p. 18):
PCV = pressure control valve.

The symbol A is often used to cover all types of analyser, but sometimes the letter representing the analyser type (e.g. D for density) is used instead.

When the general symbol A is used, the type of analyser may be specified alongside the "balloon". Other instruments (e.g. I/P converters) may be represented similarly.

Reid vapour pressure analyser.

Current-to-pressure
converter in a flow loop.

ABBREVIATIONS

LETTERS	1st POSITION	2nd POSITION	3rd POSITION
A	ANALYSER (also An)	ALARM	
B	BURNER, FLAME/FIRE		
C	CORROSION	CONTROL	CONTROLLER
D	DENSITY		
E	EMERGENCY	ELEMENT	
F	FLOW		
Fr	FLOW RATIO		
G	COMBUSTIBLE GAS	GAUGE	
H	HAND/HEAT		HIGH
HH			EXTRA HIGH
I	CURRENT	INDICATOR	
J	POWER	SCAN	
K	TIME		
L	LEVEL	INDICATOR LAMP	LOW
LL			EXTRA LOW
M	MOTOR		
N	NEEDLE		
O		OPERATED	OPEN
P	PRESSURE/PISTON	PROBE	
Pd	DIFFERENTIAL PRESSURE		
Q		TOTALIZER	
R		RECORDER	
S	SPEED/SMOKE	SWITCH	SHUT (Closed)
SO		SOLENOID OPERATED	
T	TEMPERATURE	TRANSMITTER	TRANSMITTER
U	MULTI-VARIABLE	MULTI-FUNCTION	
V	VIBRATION		VALVE
W	WEIGHT	WELL	
X	MISCELLANEOUS		
Y		CONVERTER	
Z	POSITION	SAFETY	

Introduction

In the chemical process industry it is necessary to monitor process streams continuously. This is usually done by an instrument connected on-line, and known as a process-stream analyser or simply called an analyser. Such analysers include monitors of physical parameters such as density, viscosity, vapour pressure and boiling point. Instruments that monitor other physical parameters such as flow, level, operating pressure and operating temperature are not generally regarded as analysers, however.

An alternative to the name analyser is automatic quality measuring instrument, which is often preferred, because it defines the function of the instrument.

Analysers can be classified as

(i) property-measuring analysers, such as viscometers;
(ii) composition analysers, such as chromatographs.

1.1 HISTORY

Process analysers were developed from automated laboratory equipment, modified and improved to give long life under arduous climatic conditions and safety in conditions of electrical and other hazard. The limitations of batch-type methods were early recognized, and attention was focused on continuous analysers.

In attempts to satisfy a large number of varied individual requirements, some analysers were made very complicated and were expensive and unreliable. It was then realized by users and manufacturers that such instruments were not suitable for on-line application, and more reliable analysers were developed, with emphasis on modular construction so that a wide variety of needs could be met with a standard basic instrument and suitable specialized modules. With the exploitation of microprocessors, analysers are now relatively cheaper, more versatile and more reliable, and much more widely used.

When analysers were originally applied to direct process control, many safeguards had to be incorporated to reduce the occurrence of inconvenience (or even catastrophe) when the instruments failed. Many of these safeguards are no longer necessary with modern analysers, which with good maintenance can go for a year or more without failure.

The most recent developments have been the considerable improvements in the electronic and optical components, and miniaturization, especially of some sensor elements. Sensor selectivity has been significantly improved, and the effect of some sensor-poisons reduced.

1.2 SELECTION OF THE METHOD

A method is sometimes selected for a particular measurement or analysis according to laboratory experience and custom rather than because it is the best method. This is usually because a large amount of data on the performance of the method has been collected, or because those responsible for selection of the process analyser do not want to risk financial or technical problems which might ensue if a change were made. Automation of a satisfactory laboratory method may not result in the best process analyser. For example, process boiling-point analysers have been developed which give results that correlate well with those obtained in the laboratory by use of the ASTM D86 method, and a simple vapour-pressure analyser is available which gives results correlating with those from the more complicated ASTM Reid vapour pressure analyser, and these analysers seem better suited for on-line work.

Another consideration in choosing a system is that non-specific analysers are really suitable only for determining the purity of a two-component mixture. If more than one impurity is to be determined, the analyser must give a specific response to each.

Other factors include cost, performance, safety, ease of maintenance and repair, reliability and speed.

1.3 APPLICATIONS

As well as the obvious applications to quality control and acceptance, analysers may be used for ensuring that the product can be handled economically and safely during production, blending, transportation and use.

All products must be acceptable with respect to price and quality. It is no use producing a product of too good a quality at an uncompetitive price, because nobody will buy it — unless the producer has a particularly good sales force.

Thus a control analyser on a process unit will be used to control the unit optimally or to control the utility supply rate for a given pre-determined quality value, whereas a blending or acceptance analyser will be required to control, or just monitor, the quality of the product according to the customer's requirements.

Process analysers are used for three different basic duties:

(i)   monitoring the quality of process streams;
(ii) providing a means for the operation of quality or impurity-level alarms;
(iii) process-plant quality control in order that the process unit makes an acceptable product economically.

For quality control, process analysers should be used instead of laboratory analysis if the latter is not sufficiently accurate or precise, or if very frequent and rapid measurements are required because of fairly rapid and appreciable variations in process stream quality. Process analysers may also save laboratory manpower.

Laboratory samples are usually transported in suitable containers from the sampling point, but if frequent measurements are required and the distance is not too great, a sample-transfer line may be installed between the sampling point and the laboratory. However, such continuous transfer systems are not often used, especially in large plants, because the lines are too long, or for safety reasons. An exception is sample transfer to octane-rating analysers (engines). To minimize analysis time, a process analyser should be installed as close as possible to the sampling point, to avoid time lag in getting the sample into the analyser; signal transmission to the control room can be made virtually instantaneous.

If the performance of a process analyser is superior to that of the corresponding laboratory analyser, its use for measuring product purity can decrease the quality 'give-away', i.e. the quality margin (taken as the sum of the safety margin and half the control-range error, Fig. 1.1) necessary to maintain the quality at least equal to the acceptance level of the quality specification. If the quality of the product can be made more uniform by feedback control of the process from the analyser output signal, the give-away may be still further reduced, since the safety margin depends on the accuracy of calibration and the repeatability (i.e. the peak-to-peak variation) of the measurements. The result of reduced give-away is increased output. The give-away must be kept as low as possible, consistent with guaranteed quality, since it represents an additional manufacturing cost with no equivalent return.

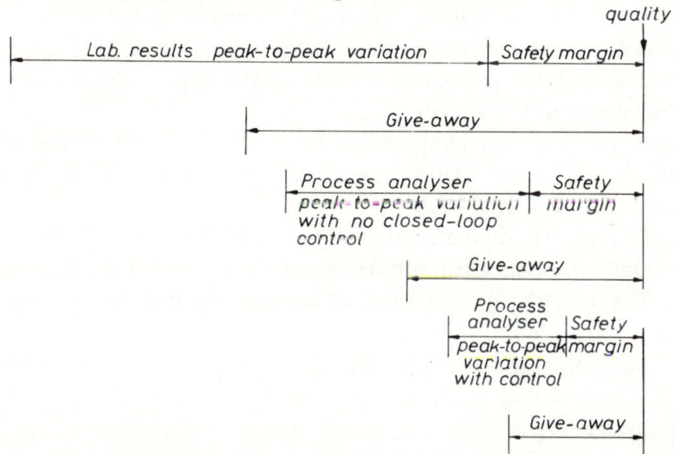

Fig. 1.1 – Relationships between quality variations, safety margins and give-away to ensure quality specifications are met.

The safety margin and peak-to-peak variation can be reduced practically to zero if a product-blending tank is used, but this may be uneconomical if sufficiently good analyser control can be achieved.

It is very important that the duties of the process and the laboratory analysers should be understood and not unnecessarily duplicated. The laboratory analyser is required for the monitoring and maintenance of quality throughout a process plant. The process analyser should be required to analyse only those streams which must be monitored for the safe and efficient control of a process unit, or for product blending and acceptance when laboratory measurements are not acceptable, or when closed-loop process-control by the analyser gives improved output at lower cost.

There are two major reasons why not all types of equipment are available in their most desirable form: one is development time and the other is cost. Most of the important research has been done, however, and it is now mainly necessary only to apply economically the methods already available, using proven high-reliability components which are safe.

Few manufacturers will build a single model of a special analyser at an economic cost, but many are interested if a reasonable number will be required. Most development nowadays is associated with attaining faster and more economical systems with improved precision, and increased reliability with minimum maintenance. The cost of maintenance is increasing rapidly relative to the initial cost of analysers, so maximum reliability and minimal maintenance are of utmost importance; unfortunately, however good their design from this point of view, analysers still have to be visited frequently to check their performance.

1.4 PREFERRED METHODS

Continuous analysers are generally preferable to automated batch-type analysers because the reliability of an analyser is generally inversely related to the number of switching operations. A continuous trend signal is often required, for example, and the additional need for a peak-picker when a batch-type analyser is used will decrease the reliability of the system.

The nearer the operating temperature and pressure are to normal atmospheric conditions the better. The system should be more reliable if regulation of these parameters is not required.

The use of reagents, diluents, carrier gases and flowing reference fluids should be minimized because of problems associated with their control and replacement. The availability and cost of special and high-purity chemicals must be considered carefully.

Self-calibrating analysers or those that do not require frequent calibration are preferable. Differential analysers have advantages provided a reference flow is not necessary.

Combustion methods should be avoided, if possible, especially if they cause oxidation or corrosion of apparatus, and deposition of combustion products.

Corrosive sample components and products should be minimized.

Analyser components in contact with the fluid to be analysed should be made from relatively inert materials, even if this is not strictly necessary, since it will increase the number of applications of the analyser and cater for abnormal conditions. The best materials are usually stainless steel for metal parts, and inert plastics such as polytetrafluoroethylene (PTFE) for plastic parts. Viton is often preferred for O-rings. The various utilities such as air and cooling water should be clean, and the air should be sufficiently dry.

1.5 ORGANIZATION OF THE HANDBOOK

Analysers are usually fairly complex pieces of apparatus, and in some senses involve chemical engineering in miniature. For convenience, this work is divided into separate volumes based mainly on the type or principle of the measurements made. Because of the nature of the applications of analysers, this first volume begins with some definitions and conventions that will be repeatedly used in the handbook and are in common use in industry, before going on to consider general aspects of analysers. The remaining four volumes deal with specific types of analyser in detail.

CHAPTER 2

Units and symbols

2.1 UNITS

Though scientists work practically exclusively with metric units and a determined effort is being made to ensure that only SI units are used, in many industries and in common parlance in the Anglo-Saxon world, other units are still in use. Moreover, excellent though SI units may be for theory, they are often rather remote from what is measured in practice. Because this handbook is intended for practical use in industry, the units used will be the most appropriate for the purpose, but to avoid needless difficulty in interpretation we shall give a brief account of some of the more confusing terminology and interconversions, especially in the case of pressure, which is the outstanding example of the gulf between theoretical (SI) units and practical measurement units. A multiplication sign ($\cdot$) or a slash will be used between components of combination units, as appropriate, to eliminate possible confusion, between m for metre and m for milli.

Force

$$F = ma$$

A force F of 1 newton (N) will give a mass m of 1 kg an acceleration a of 1 m.sec^{-2}. Thus 1 newton = 1 kg.m.sec^{-2}.

Energy

$$\text{Work} = \text{force x distance}$$

$$E = Fd$$

$$1 \text{ joule (J)} = 1 \text{ N.m}$$

Another form of energy is thermal energy, or heat (Q); since the introduction of SI units, the calorie has been replaced by the joule. We therefore refer to the heat content of a fluid in terms of kJ/kg.

The total energy of a flowing liquid is the sum of its internal energy, kinetic energy due to movement, and potential energy due to position. For a mass m kg with internal energy u kJ/kg, moving at c m/sec at a height h m, the total energy is

$$E = \text{I.E.} + \text{K.E.} + \text{P.E.}$$

$$= mu + \frac{mc^2}{2} + mgh \text{ kJ}$$

where g is the acceleration due to gravity, in m.sec^{-2}.

Power

$$W = \frac{\mathrm{d}E}{\mathrm{d}t} \text{ watts}$$

A power of 1 watt (W) is work done at the rate of 1 J/sec. In electrical systems the power in watts is equal to the product of the current in amperes (A) and the voltage in volts (V).

Pressure

Pressure is the force per unit area.

$$P = \frac{F}{A}$$

The units used for pressure are many and varied, and require a distinction to be made between absolute pressure and the pressure measured on a gauge (i.e. relative to atmospheric pressure). As far as we are aware, the SI system totally fails to take account of the fact that gauge pressures are commonly used in practice. We therefore regard it as essential to retain the working units of pressure such as psig (pounds force per square inch, gauge pressure) or kg/cm^{2}g, at least for ready reference when the chemical plant being monitored has all its gauges and pressure recorders calibrated or graduated in these units. A pressure quoted without the suffix g then means an absolute pressure.

The following units have been used at various times and in various countries.

ATA   absolute technical atmosphere (1 kgf/cm^2)

ATU   gauge technical atmosphere, U standing for the German *Überdruck* (excess pressure, i.e. above atmospheric pressure).

AT   technical atmosphere differential pressure.

kp/cm^{2}g   kilopond force per square cm, gauge pressure (this is the same unit as the ATU).

bar   1.01325 ATA [= 10^5 N/m^2 = 10^5 Pa (pascals)] .

The Germans have used the ATA and ATU. The French have used the bar for some time and this is becoming generally used in industry. The American Society for Testing and Materials (ASTM) recommended the use of MN/m^2 (i.e. 10^6 N/m^2) but has now changed to use of the pascal (N/m^2), the SI unit (recommended by the Germans in DIN 1314, December 1971).

Before metrification the British and Americans used psia (pounds force per square inch absolute), psig (for gauge pressures), and psi (or psid) for differential pressure, and many still do.

For low gauge and differential pressures, the units commonly used are mmH_2O, in WG or ft WG. For measurements approaching atmospheric pressure and for sub-atmospheric pressure (vacuum), the units commonly used are mmH_2O at $0°C$ or torr (1 atm = 760 torr) and mmHg. When columns of water or mercury are used, then the reference temperature must also be stated because of its effect on density.

Scientists commonly use the physical standard atmosphere (equal to 1.0332 absolute technical atmospheres). The difference between the 'physical' and 'technical' pressure units is that the former are based on mass and the latter on weight, the conversion factor being given by the gravitational acceleration and the units of area.

In this handbook the **bar** will generally be used as the unit of pressure, but other units may be used if more suitable for a particular purpose.

If it is necessary to convert gauge pressure accurately from one unit to another it is often desirable to do the conversion in units of absolute pressure. The relationship between absolute and gauge pressure is shown diagramatically below:

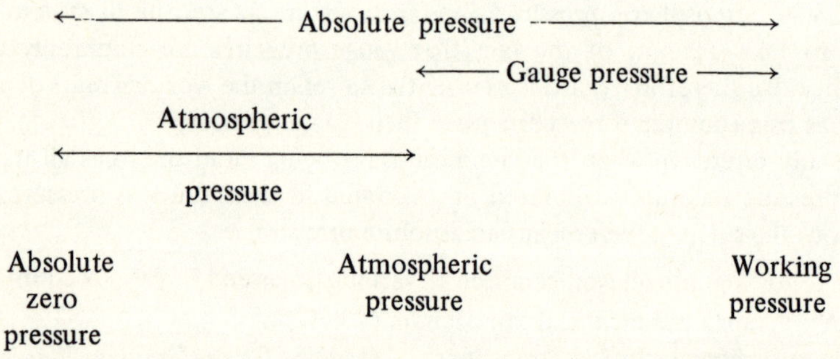

If an American refers to a pressure in psi, he probably means psig. Likewise a pressure stated in bars is often meant to be a gauge pressure. It is for these reasons that the retention of the g for gauge is strongly recommended, to avoid all ambiguity (especially in view of the lack of SI equivalents for gauge pressures).

Reid vapour pressure is often expressed in lb, which should be interpreted as meaning psi absolute.

The relationships between the units of pressure above are given in the following table:

	Absolute pressure		*Gauge pressure*
1 atm =	std atm at sea level		
=	1·0332 kgf/cm² abs	1 atm =	1 kgf/cm² g
=	1·0332 kp/cm² abs	=	1 kp/cm² g
=	1·0332 ATA	=	1 ATU
=	1·01325 × 10⁶ dynes/cm² abs		
=	1·01325 bar abs	=	0·980665 bar ga
=	1·01325 × 10⁵ N/m² abs	=	0·980665 × 10⁵ N/m² ga
=	0·101325 MN/m² abs	=	0·0980665 MN/m² ga
=	1·01325 × 10⁵ Pa abs	=	0·980665 × 10⁵ Pa ga
=	760 mmHg at 0°C		
=	760 torr		
		=	10 000 mm H₂O at 4°C
		=	10 018 mm H₂O at 20°C
=	14·696 lbf/in² abs (psia)	=	14·2234 lbf/in² g (psig)

Note that ga is used instead of g to indicate gauge when SI units are used, to avoid confusion with g for gram.

Pressure Rating

The pressure rating of a component is often expressed in lb, which is the maximum allowable pressure in psig at an elevated temperature.

The required acceptance test pressure for the component is often specified as 1.5 times the maximum allowable working pressure of the equipment at some agreed temperature such as 100°F (38°C).

Volume

This is expressed in litres (l.), now defined as 1 litre = 1 dm³, or in m³.

$$1 \text{ m}^3 = 10^3 \text{ litres}$$

The specific volume of a fluid is the volume occupied by unit mass of the substance and is the reciprocal of the density. The unit commonly used is m³/kg.

Temperature

The kelvin is defined as the unit of temperature determined by the Carnot cycle with the triple-point temperature of water defined as exactly 273.16 K. The International Practical Celsius Scale is defined by a set of equations based on selected reference temperatures, the most familiar being the boiling point of pure water at 1 'physical' std atm pressure (100°C), and the freezing point of pure water at 1 std atm (0°C). The degree Celsius (°C, formerly — and historically more correctly — called the degree centigrade) has the same value as the kelvin for defining a temperature *difference*, but the two scales have different reference zero, and are related by adding 273.15 to the kelvin temperature to obtain the Celsius temperature. The Celsius scale is the one commonly used.

2.2 REFERENCE CONDITIONS

Because of the gas laws, when volumes of gases are involved in calculations it is necessary to specify the relevant temperatures and pressures. In the past a great deal of confusion has arisen from the use of the terms standard temperature and pressure (STP) and normal temperature and pressure (NTP) more or less interchangeably and with different values used for the standard temperature (although the standard pressure has always remained 1 'physical' std atm).

The reference temperatures commonly used have been variously 0, 15, 20 and 25°C. The British Standard BS 3379:1978, based on the International Standards Organization standard ISO 5024, defines standard reference conditions for measurement of petroleum liquids and gases, these being a temperature of 15°C and a pressure of 1 standard atmosphere. As other reference conditions may be used in industries other than the petroleum industry, we propose to indicate the reference temperature (in °C) by a numerical subscript to the relevant symbol. For example, V_{20} would mean a volume measured at 20°C and 1 std atm, or 26 m^3_{20} would mean 26 m^3 measured at 20°C and 1 std atm.

2.3 SYMBOLS

The symbols used for physical quantities and units will generally be those recommended by IUPAC and readily recognizable, or will be defined in the text. The prefixes used for multiples and sub-multiples of units are listed below.

Factor	Prefix	Symbol
10^{12}	tera	T
10^{9}	giga	G
10^{6}	mega	M
10^{3}	kilo	k
10^{-1}	deci	d
10^{-2}	centi	c
10^{-3}	milli	m
10^{-6}	micro	μ
10^{-9}	nano	n
10^{-12}	pico	p

Numerous diagrams showing analyser systems are given in the book, and standard parts such as valves, filters and so on are indicated by formal symbols, shown on pp. 11–17.

CHAPTER 3

Concentration of components of mixtures

3.1 GAS MIXTURES

The concentration of a component of a gas mixture may be expressed in several ways, such as weight fraction (w/w) or percentage, volume fraction (v/v) or percentage, mole fraction, parts per thousand or per million (w/w, w/v, v/v), etc. For concentrations below the parts per million (ppm) level, it is common practice in the United States to use the term ppb (meaning parts per billion, which in this case is the American billion, 10^9). This usage is confusing, because the European billion is 10^{12}, and moreover very few authors indicate whether they are referring to w/w, w/v or v/v concentrations. It is far better to express the concentration in appropriate units such as ng/g, ng/ml, ml/m^3, which convey all the necessary information. The use of expressions such as pptm (for parts per ten thousand million) should not be countenanced, since the meaning is not obvious.

Concentration by weight
The concentration of the ith component of a mixture, expressed in % w/w, is clearly

$$\% \, \text{w/w} = \frac{w_i}{\Sigma w_i} \times 100$$

where w_i is the weight of the ith component and Σw_i is the sum of the weights of all the components. For lower concentrations, we can use μg/g as the units,

$$\mu\text{g/g} = \% \, \text{w/w} \times 10^6/10^2 = 10^4 \times \% \, \text{w/w}$$

Concentration by volume
The concentration in % v/v, for component i, is analogously

$$\% \, \text{v/v} = \frac{V_i}{\Sigma V_i} \times 100$$

where V_i is the volume of the ith component, and hence

$$\mu 1/1. = \% \, v/v \times 10^6/10^2 = 10^4 \times \% \, v/v$$

The composition will refer to a particular temperature and pressure, and for real gases may change with the conditions of measurement because the compressibility varies from gas to gas. For n moles of a given gas having a compressibility factor Z, we can write

$$PV = ZnRT$$

where P is the absolute pressure, V the volume, R the gas constant and T the absolute temperature (all in compatible units). If P is in atm and V in litres, R has the value 0.08206 1.atm.K^{-1}.mole^{-1}, whereas if P is in Pa and V in m^3, R is 8.3143 N.m.K^{-1}.mole^{-1} = 8.3143 J.K^{-1}.mole^{-1}. The **mole** has been defined in the SI system as a base unit for the physical quantity called 'amount of substance' and is the 'amount of substance' that contains as many elementary entities (which *must* be specified) as there are carbon atoms in 0.012 kg of carbon-12. For practical purposes the mole is simply Avogadro's number of the entities specified (which may be atoms, molecules, ions, electrons, particles, groups etc.).

Thus for a mixture of gases at a given temperature and pressure, the concentration by volume for each component is given by

$$\% \, v/v = \frac{V_i}{\Sigma V_i} \times 100$$

$$= \frac{n_i Z_i RT/P}{\Sigma n_i Z_i RT/P} \times 100$$

$$= \frac{n_i Z_i}{\Sigma n_i Z_i} \times 100$$

$$= \frac{y_i Z_i}{\Sigma Z_i} \times 100$$

where y_i is the molar fraction of the ith component of the mixture.

$$y_i = n_i/\Sigma n_i$$

The molar fraction can also be expressed as mole % concentration:

$$\text{mole} \, \% = \frac{n_i}{\Sigma n_i} \times 100 = y_i \times 100$$

Parameters related to concentration
From Dalton's law of partial pressures, the **total pressure** will be the sum of the

partial pressures:

$$P = \Sigma P_i$$

$$= \frac{RT}{V} \Sigma Z_i\, n_i$$

The **density**, ρ, is the weight per unit volume:

$$\rho = w/V \text{ kg/m}^3 \quad (\text{or g/l.})$$

For an ideal gas, the density at $0°C$ and 1.01325 bar, which we will call the reference density (ρ_r), is given by

$$\rho_r = \frac{(MW)}{22.4} \text{ kg/m}^3$$

where (MW) is the weight of 1 mole of the gas (the 'molecular weight'). To obtain the working density from this reference density, it is necessary to correct for temperature and pressure. For real gases it is also necessary to apply a correction for the compressibility. Thus the volume of 1 mole of a real gas at $0°C$ and 1.01324 bar is $22.4\, Z_i$ litres, where the value of Z_i is that for $0°C$ and 1.01325 bar. Thus the density at temperature T (kelvin) and pressure P (bar), $\rho_{T,P}$, is

$$\rho_{T,P} = \frac{(MW)}{22.4\, Z} \times \frac{273}{T} \times \frac{P}{1.01325}$$

$$= 12.03\ (MW)\, P/TZ \text{ kg/m}^3$$

As industry is generally dealing with mixtures of gases, it is convenient to have a uniform scale of comparison between individual gases, and this can be achieved by using the so-called **reduced variables**, X_R, obtained by dividing the observed value X by the critical value X_c (obtained from van der Waals isotherms [1]):

$$\text{Reduced pressure } P_R - P/P_c = 1.01325/P_c\ (P_c \text{ in bar})$$

$$\text{Reduced temperature } T_R = T/T_R = 273.16/T_c\ (T_c \text{ in kelvin})$$

$$\text{Reduced volume } V_R = V/V_c = 22.4/V_c\ (V_c \text{ in l./mole})$$

$$\text{Reduced compressibility factor } Z_R = P_R\, V_R/RT_R$$

Table 3.1 gives the values of reduced variables and compressibility coefficients for some gases at $0°C$ and 1.01325 bar.

Table 3.1

Reduced variables and the compressibility factors of some common gases at 0°C and 1.01325 bar

Gas	P_c	P_R	T_c	T_R	Z	$22.4\,Z$
Oxygen	51.4	0.020	154	1.77	0.999	22.38
Hydrogen	21.5	0.048	41	6.6	1.000	22.40
Methane	47.3	0.022	191	1.43	0.977	22.33
Ethane	50.4	0.021	305	0.89	0.989	22.15
Ethylene	52.6	0.020	283	0.97	0.992	22.22
Propane	43.4	0.024	370	0.74	0.975	21.84
Propylene	46.5	0.022	366	0.75	0.978	21.91
Isobutane	38.2	0.027	407	0.67	0.953	21.35
n-Butane	37.2	0.028	426	0.64	0.935	20.92

The weight (kg) of 1 m³ of a gas at temperature T (kelvin) and pressure P (bar) will be

$$w = 12.03 \, (MW) \, P/TZ \text{ kg}$$

For a mixture of gases, the **weight % concentration** of the ith component, already defined as $100 \times w_i / \Sigma w_i$, can also be expressed as

$$\% \text{ w/w} = \frac{(MW)_i}{(\overline{MW})} \times \% \text{ v/v}$$

where $(\overline{MW})$ is the **mean molecular weight** of the mixture:

$$(\overline{MW}) = \Sigma y_i \, (MW)_i$$

The **specific gravity** of a gas, g, is defined (for metric units) as its density relative to that of air at 0°C and 1.01325 bar abs (1.293 kg/m³).

$$g = \rho_{gas}^{T,P} / \rho_{air}^{0,\,1.01325} = \rho_{gas}^{T,P} / 1.293$$

The **relative density** of a gas, γ, is defined as the density relative to that of air at the same temperature and pressure:

$$\gamma = \frac{\rho_{gas}^{T,P}}{\rho_{air}^{T,P}} = \frac{(MW)_{gas}}{(MW)_{air}} = \frac{(MW)_{gas}}{28.97}$$

Since

$$\rho_{air}^{T,P} = \rho_{air}^{0,\,1.01325} \times \frac{P}{1.01325} \times \frac{273}{T}$$

$$\gamma = \rho_{gas}^{T,P} / \rho_{air}^{T,P} = g \times \frac{1.01325}{P} \times \frac{T}{273}$$

The sample presented to an analyser must be a single-phase sample. There will be a marked effect on a gas analysis if a small proportion of one component is present in the liquid phase. For a given amount of a component, w (kg), the volume (m^3) in the liquid phase would be

$$V_L \;=\; w/\rho_L$$

and in the vapour phase

$$V_V \;=\; w/\rho_V^{T,\,P}$$

$$= wTZ/12.03 \,(MW)P$$

Hence the expansion that occurs on vaporization of a liquid sample is seen to be

$$\frac{V_V}{V_L} \;=\; 0.0831 \, \rho_L TZ/(MW)P$$

Example: For isobutane, $\rho_L = 566$ kg/m^3, (MW) $= 58.1$ and $Z_R = 0.953$. Hence at 20°C and $P = 2$ bar abs,

$$\frac{V_V}{V_L} \;-\; 0.0831 \times 566 \times 293 \times 0.953/58.1 \times 2.0265$$

$$\sim \; 112$$

This ignores the effect of temperature on the density of the liquid, but this is negligible in comparison with the gas/liquid volume-ratio.

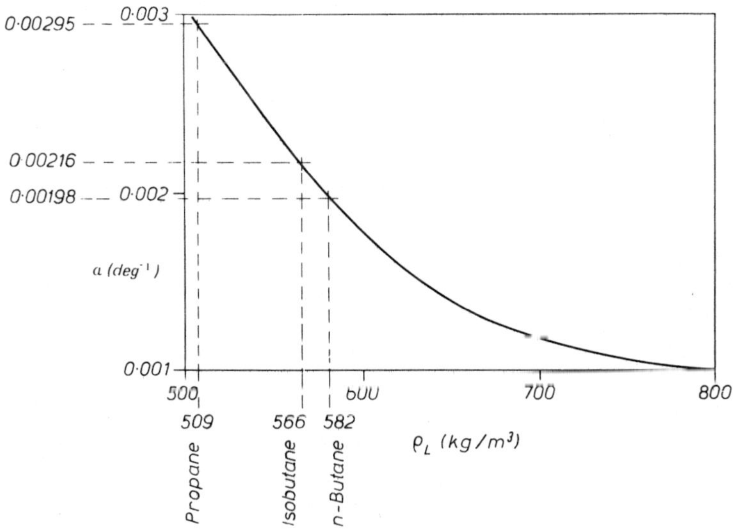

Fig. 3.1 – Variation of temperature coefficient of volume expansion of a liquid with density.

The effect of temperature on the density of the liquid can be taken into account by using α, the coefficient of cubical expansion at constant pressure:

$$V_2 = V_1(1 + \alpha \Delta T)$$

where V_2 is the volume at a temperature ΔT (°C) different from that at which V_1 was measured. This gives

$$\rho_2 = \frac{w}{V_2} = \frac{\rho_1 V_1}{V_2} = \rho_1/(1 + \alpha \Delta T)$$

A useful working rule is that $\alpha \sim 0.36/(\rho_1 - 400)$ for paraffinic hydrocarbons with densities of 500–900 kg/m^3 at room temperature.

If the temperature approaches the boiling point of the hydrocarbon at the particular working pressure, the density drops more rapidly than is indicated by this linear approximation (Fig. 3.1). This applies to propane and the lighter paraffinic hydrocarbons at normal ambient temperatures. Thus, when lighter paraffinic components are present, the density will decrease more rapidly, according to the concentration of these components, as their boiling points are approached.

3.2 CONCENTRATION OF COMPONENTS IN LIQUID MIXTURES

A solution consists of one or more solutes dissolved in a solvent. The concentration of a component of the solution can be expressed in several ways.

The **mole fraction** (x_A) of component A is the number of moles of component A divided by the total number of moles in the system ($x_A = n_A/\Sigma n$).

The **molarity** (M), or **amount-of-substance concentration**, is the number of moles of component A per litre of solution, or the number of kmoles per m^3.

The **molality** (m) is the number of moles of solute in 1 kg of solvent, or the number of kmoles of solute per 1000 kg of solvent. The **volume molality** (m') is the number of moles of solute per litre of solvent.

The **titre** is the reacting strength of a standard solution. It is usually expressed as the weight of titrated substance equivalent to 1 ml of the standard solution. For example, a solution with a titre of 1 mg of chloride contains in each ml an amount of a substance that will react quantitatively with 1 mg of chloride.

The **equivalent** of a substance is that amount of substance which will provide or react quantitatively with 1 mole of hydrogen ions in a neutralization reaction, or 1 mole of electrons in a redox reaction, or react with one equivalent of any other species. When a reaction can give rise to more than one product, depending on the conditions used, the reaction must be specified.

The **normality** is the number of equivalents of solute per litre of solution. Equinormal solutions contain equal numbers of equivalents per unit volume, so equal volumes will react exactly with each other.

Example: A solution contains W g of sulphuric acid, H_2SO_4, per litre. The molarity of the solution, C, is equal to $W/98.08$. The molarity of SO_4^{2-} is also C, but of H^+ is $2C$. One mole of H_2SO_4 will provide two moles of hydrogen ions, so the equivalent is equal to half a mole, i.e. 49.04 g. Thus, the normality of the solution is $W/49.04 = 2C$.

The relationships between the equivalent and the mole for some species and reactions are shown in Table 3.2.

Table 3.2

Solute	Reaction	Equivalent
NaOH	$OH^- + H^+ \longrightarrow H_2O$	1 mole
$Ba(OH)_2$	$2OH^- + 2H^+ \longrightarrow 2H_2O$	mole/2
$FeSO_4$	$Fe^{2+} \longrightarrow Fe^{3+} + e^-$	1 mole
$Na_2(COO)_2$	$(COO)_2^{2-} \longrightarrow 2CO_2 + 2e^-$	mole/2
$KMnO_4$	$MnO_4^- + 5\,e^- + 8H^+ \longrightarrow Mn^{2+} + 4H_2O$	mole/5
$KMnO_4$	$MnO_4^- + 3\,e^- + 4H^+ \longrightarrow MnO_2 + 2H_2O$	mole/3
$KMnO_4$	$MnO_4^- + e^- \longrightarrow MnO_4^{2}$	1 mole

Interconversion between molality (m) and molarity (M) is done as follows. Suppose that b g of solute B [molecular weight $(MW)_B$] are dissolved in a kg of solvent A to obtain V litres of solution of density ρ_s kg/l. Then the following relations hold.

$$V \;\; = (a + 0.001b)/\rho_s$$

$$M_B = b/(MW)_B V$$

$$m_B = b/(MW)_B a$$

Hence

$$M_B = b\,\rho_s/(a + 0.001b)\,(MW)_B \;\; = am_B\,\rho_s/(a + 0.001b)$$

and

$$m_B = M_B\,(a + 0.001b)/a\,\rho_s$$

The molality is also related to the mole fraction x, since 1 kg of solvent A is $1000/(MW)_A$ moles of A, $(MW)_A$ being the molecular weight of A. Hence, if n_A and n_B are the numbers of moles of A and B respectively:

$$x_B = \frac{n_B}{n_A + n_B} = \frac{b/(MW)_B}{[1000a/(MW)_A] + [b/(MW)_B]}$$

$$= \frac{am_B}{[1000a/(MW)_B] + am_B}$$

Hence

$$m_B = \frac{1000\,x_B}{(MW)_B(1-x_B)}$$

REFERENCES

[1] P. W. Atkins, *Physical Chemistry*, Oxford University Press, Oxford, 1978, p. 44.

CHAPTER 4

Errors, calibration and correlation

4.1 ERRORS

The result of any analysis, and therefore the signal from an analyser, is subject to many errors. These may be classified as random, systematic and gross.

Random errors result from the impossibility of having complete control over all the experimental variables. They are treated by the methods of statistics. The applications of statistics to analytical results have been described in several texts and articles [1-11].

Systematic errors cause incorrect results to be obtained: they can usually be assigned to definite causes. They are normally unidirectional, and they may be constant (e.g. resulting from impure reagents, insufficient selectivity, etc.) or proportional to the analyte concentration (e.g. resulting from incorrect assumption of linearity, see Fig. 4.1, or from a difference between the calibration slopes for sample and standard). They may vary with time, sometimes cyclically. The total systematic error of a procedure is called the bias. Systematic errors can be reduced by calibration, e.g. with standard reference fluids.

Gross errors generally result from mistakes or 'catastrophic' events.

The **absolute error** e_a of a particular measurement x is given by

$$e_a = x - x_t$$

where x_t is the true result.

The **relative error** e_r is given by

$$e_t = \frac{x - x_t}{x_t}$$

This is frequently expressed as a percentage

$$e_r (\%) = \frac{x - x_t}{x_t} \times 100\%$$

The value of x_t is generally not known, and it is replaced in these equations by the mean, $\bar{x}$, of a number of readings. This is a reasonably good estimate of x_t, provided that systematic errors have been eliminated or corrected for.

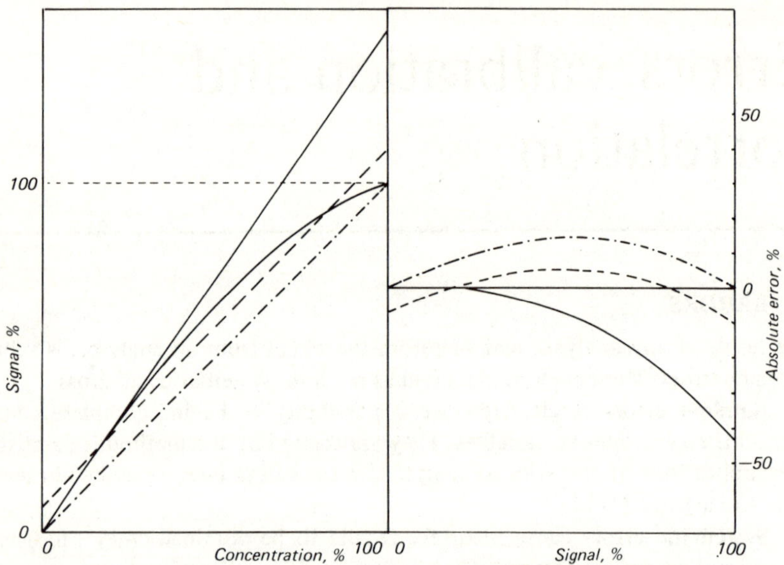

Fig. 4.1 – The effect of incorrect assumption of linearity, e.g. in the measurement of concentration by optical absorption methods – the Beer–Lambert law is frequently not obeyed at high concentrations.

An estimate of the **systematic error**, or **bias**, is given by

$$e_s = \bar{x} - x_t$$

The **terminal error** is the error at 100% quality. (Most analysers cannot measure 0% and 100% quality. It is often easier to measure the concentrations of the impurities near to 0% than to measure the purity of a substance that is nearly 100% pure.)

The **accuracy** of a result (or set of results) is its nearness to the true value. For a single result, the accuracy is characterized by the absolute error, $x - x_t$; for a set of results by $\bar{x} - x_t$. The accuracy of an instrument is often specified relative to the full-scale deflection, if it has a linear response. If it has a non-linear response (e.g. logarithmic), the accuracy is often specified relative to the value measured, down to a certain minimum value. In manufacturers' specifications, definitions of accuracy will vary according to the type of analyser and its intended field of use.

The **precision** of a set of results refers to the scatter or dispersion of the values about the central value. Provided that the values follow the *normal* (or *Gaussian*) *distribution*, at least approximately, accuracy is best characterized by the standard deviation, σ. However, σ refers to an infinite number of values;

for a set of n values, an estimate, s, of the value of σ is given by

$$s = \sqrt{\frac{\Sigma (x - \bar{x})^2}{(n-1)}}$$

However, if n is less than 10, s is better determined from the *range* of the values (the difference between the largest and smallest) according to

$$s = kR$$

The values of k are given in Table 4.1.

Table 4.1

n	k	n	k
2	0.8862		
3	0.5908	7	0.3698
4	0.4857	8	0.3512
5	0.4299	9	0.3367
6	0.3946	10	0.3249

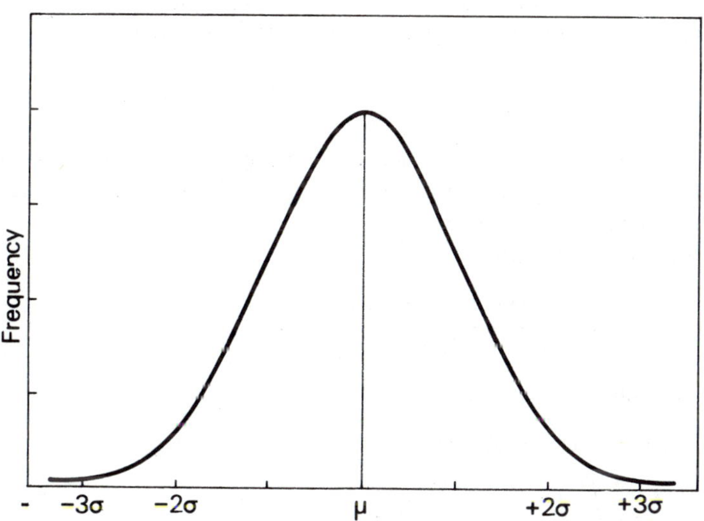

Fig. 4.2 – The normal (Gaussian) distribution.

The best way to use the standard deviation to give a measure of the precision is to quote **confidence limits**. For a normal distribution, shown in Fig. 4.2, it can be shown mathematically that the area under the curve from $(\mu - \sigma)$ to $(\mu + \sigma)$ is about 68.3% of the total area, from $(\mu - 2\sigma)$ to $(\mu + 2\sigma)$ about 95.5%, and from $(\mu - 3\sigma)$ to $(\mu + 3\sigma)$ about 99.7%. Thus, in a large set of values, about 2/3 will lie within $\pm\sigma$ of the population mean, μ (which is approximately equal to its estimate, the sample mean, $\bar{x}$), 95.5%, or about 19 in 20, will lie within $\mu \pm 2\sigma$, and 99.7% will lie within $\mu \pm 3\sigma$. Thus the results of an analysis may be reported as equal to $\bar{x} \pm 2s$ (95% confidence limits).

If the number of data points is small (< 30) the normal distribution should be replaced by the 'student' t distribution, which has a greater frequency of larger errors and a smaller frequency of smaller errors. The confidence interval for the *mean* is then $\mu = \bar{x} \pm ts/\sqrt{m}$, where the value of t is a function of n and of the desired degree of probability. Tables can be found in any statistics textbook (e.g. [5–7]).

In work with analysers, two additional terms related to precision are commonly used. The **repeatability** is the precision of a single analyser for a single sample, when used by the same operator. The measurements should be made at regular intervals of about one hour over a period of about 100 hours, although a period of as little as 30 hours may sometimes be acceptable. At the end of the test period, the drift should be noted.

The **reproducibility** is the precision for a number of different analysers, probably in different locations and with different operators, but for the same sample. The errors between analysers are due to both the analytical and sampling methods.

Repeatability and reproducibility are usually quoted in terms of 95% confidence limits.

The random errors associated with different parts of an analysis may reinforce one another, or cancel in whole or in part. The rules for calculating the overall random error are as follows:

(1) For addition or subtraction:

$$y = K + a_1 x_1 + a_2 x_2 - a_3 x_3 + \ldots$$

$$s_y = \sqrt{a_1^2 s_{x_1}^2 + a_2^2 s_{x_2}^2 + a_3^2 s_{x_3}^2 + \ldots}$$

(2) For multiplication or division:

$$y = k x_1^{a_1} \times x_2^{a_2}/x_3^{a_3} \times \ldots\ldots$$

$$\frac{s_y}{y} = \sqrt{a_1^2\left(\frac{s_{x_1}}{x_1}\right)^2 + a_2^2\left(\frac{s_{x_2}}{x_2}\right)^2 + a_3^2\left(\frac{s_{x_3}}{x_3}\right)^2 + \ldots}$$

(3) For logarithmic expressions:

$$y = K \ln x \qquad\qquad y = K \log_{10} x$$

$$s_y = K \frac{s_x}{x} \qquad\qquad s_y = \frac{K}{2.3} \cdot \frac{s_x}{x}$$

These equations are called the *rules of propagation of errors*.

The **limit of detection** of an analyser system depends on the method and on the equipment used. With electronic equipment alone, the limit of detection is often taken to be equal to the noise level, i.e. a signal is accepted if its amplitude is equal to or greater then the standard deviation of the noise signal. However, this definition is not satisfactory for use with analytical measurements.

The **analytical limit of detection** is defined as the minimum amount of a component that can be detected with certainty by a single analytical measurement. In practice, it is necessary to specify the degree of certainty required. To do this, two probabilities must be considered: (1) the probability of false detection, and (2) the probability of correct detection [7]. If the probability of false detection is chosen to be 0.1% and the probability of correct detection to be 99.8% then the detection limit should be taken as equal to six times the standard deviation (s_b) of the background noise. If, as sometimes occurs, it is taken as $3s_b$, there is only a 50% probability that the signal is definitely *not* from the background.

If, instead of only one measurement, n measurements are made, the detection limit is decreased to $6s_b/\sqrt{n}$. Since process analysers make many measurements in a small period of time, a limit of detection lower than $6s_b$ can usually be accepted with confidence. However, if the quality being measured by the analyser is changing very rapidly, there will be relatively few measurements for a particular value of the signal, so the limit of detection will again be nearer $6s_b$. Fortunately, most of the variations in quality that have to be measured by analysers are relatively slow, so this effect can normally be ignored.

The **resolution** of an analyser is defined as the least change in the measured quality value that can be detected with a specific degree of certainty by the user of the analyser under specified conditions in a specified time.

The **threshold** of an analyser is the smallest input that will cause a detectable output change from zero.

Selectivity is a measure of the extent to which an analyser is capable of differentiating between the component of interest and other components. It may be governed by two factors: (*i*) the capability of the analyser to separate the component of interest physically or figuratively (e.g. by wavelength selection in spectrometry) from the other components and (*ii*) the difference in sensitivity of response to the component of interest and to other components. During acceptance tests at the vendor's works, the vendor should measure the sensitivity of the instrument to all of the fluids listed in the stream composition, unless otherwise specified, and state the concentration of a given interfering impurity that would corrrespond to less than a specified fraction of the analyser range, for instance, that 10 mole % of impurity A would correspond to less than 0.5%

of the range of the analyser for the component to be determined.

The **signal range** of an analyser correponds to the input quality range, and the difference between the maximum and minimum values is the **span** of the analyser. Errors may be defined relative to the actual quality value, the maximum quality value, or the analyser span.

When the analyser response is non-linear but the control system is based on an assumed linearity, a least-squares procedure can be used to find the straight line giving least average absolute error, a line can be drawn joining the ends of the analyser span or a tangent may be drawn to the initial slope of the curve (see Fig. 4.1). It would be better to use a least-squares fit to the quality specification range, or a dedicated microcomputer with a polynomial fitting program.

4.2 CALIBRATION FLUIDS

All analysers should be *correlated* (see p. 43), but only some need a check *calibration*, and then only at a frequency depending on the rate of drift of the zero and sensitivity, and on the confidence in the analyser repeatability.

Manual injection of calibration fluids is satisfactory for infrequent checking, but automatic timed injection is better if frequent checking is needed. When the injection is manual, the zero and sensitivity should not be adjusted unless sufficient of the measured values lie outside the control limits (see p. 45). If an adjustment has to be made, the correlation must be repeated.

Some analysers have automatic zeroing and sensitivity adjustment either directly from measurement of the quality of the calibration fluid or from a reference voltage which may be checked and adjusted during analyser calibration.

If calibration fluids are stored, it is necessary to know and to check the rate of change of the quality of the fluid, so that it is not stored for too long. Quality drifts of calibration fluids can be due to changes caused by settling, reaction, diffusion or adsorption.

Contamination of the sample by the calibration fluid can occur because of leaking valves; stream-selecting ball valves are preferable to other types. Double block and bleed valve systems reduce such contamination considerably (p. 122).

Calibration gases may be stored in horizontally mounted cylinders which can be shaken. More often they are mounted vertically. Generally, gases are kept well mixed by thermal currents.

Calibration liquids should be stored in cylinders with direct gas pressurization or with indirect pressurization through use of a suitable flexible membrane. The Flamco Airfix vessel is often suitable for the latter purpose. In most cases it is necessary to maintain the cylinder pressure above the vapour pressure to prevent flashing.

Two types of calibration fluid may be used: 'zero fluid' for checking the analyser zero and 'span fluid' for checking the analyser span or sensitivity. Zero fluid contains none of the component of interest but has a quality which should correspond with the analyser zero. Span fluid should have a quality which corresponds to approximately 70% of the analyser span.

Clean air is suitable for use as the span fluid for a flue-gas oxygen analyser because it contains a sufficiently accurately known oxygen concentration.

The use of calibration fluids is further described in Chapter 6.

4.3 CALIBRATION

The process of **calibration** involves using the analyser to measure the quality of a prepared sample, of accurately known quality, which has been carefully checked in the laboratory.

When the analyser is being tested in the factory, the manufacturer should calibrate it at several points, typically near to 20, 50, 70 and 90% of analyser range. This is the initial calibration.

During operation of the analyser in the field, calibration should be checked at about 70% of analyser range, at intervals of time determined by the analyser drift and the accuracy required. When an unacceptably large drift occurs, then the analyser sensitivity may be adjusted to minimize the error. A calibration or correlation check must then be done at several points to prove that the analyser drift has been corrected satisfactorily. An unaccountable drift should always be investigated to determine its source and cause, in case maintenance is necessary. The drift may result in a change in slope of the analyser sensitivity curve and/or in a change in bias of the curve.

The calibration curve corrects for systematic errors of the analyser, but not for those due to changes in temperature and pressure. Separate curves or tables are used for these corrections.

4.4 CORRELATION AND CONTROL CHECKS

The process of **correlation** involves the comparison of two sets of measurements, one made by the process analyser and the other resulting from the laboratory analysis of samples taken from the process stream at the time the analyser measurement was taken. Shortly after an analyser is installed and has undergone acceptance tests, an extensive correlation should be run.

The accuracy obtainable with correlation is normally rather less than with calibration, because of the possibility of errors in the taking of the laboratory sample, and also in the laboratory analysis. An additional error arises because the laboratory sample may not be quite identical with the sample seen by the analyser.

The values of the component concentration measured during the correlation run must generally have a limited range because it would be uneconomic to adjust the plant to produce samples with a wider quality range than is normal. Most measurements should be between 50 and 70% of full scan, if the analyser range has been selected properly. However, this is not always the case; sometimes a range may be selected so that possible abnormal (and undesirable) concentrations of an impurity can be measured, in which case the normal low level would give a negligible signal (this is equivalent to using a 'go/no-go' gauge).

4.4.1 Procedure

Before any sample is taken, it should be ensured that the analyser reading has been stable for at least 15 min (in the case of continuous analysers) or for at least 5–10 samples (for discontinuous or sampling analysers). The laboratory sample must be taken from a location as near as possible to the sampling point of the process analyser. At the moment the sample is taken, a mark should be made on the analyser chart; a time correction should then be applied to take account of the time taken for the process fluid to travel between the two sampling points, and the total lag time of the analyser.

The process analyser readings (x) and, later, the corresponding laboratory results (x_r) are noted in a log book, along with the time of the measurement and the date. The results should not be rounded off.

4.4.2 Preparation of the Control Chart

Pairs of values of x and x_r are plotted on a graph of analyser readings *vs.* laboratory results, then the equation of the 'best' straight line through the points is determined. The equation of the line takes the form

$$x_r = mx + C$$

where m is the slope of the line ($= \tan \theta$), and C is the intercept (see Fig. 4.3). Provided that it is reasonable to assume that the laboratory results are more accurate and precise than the analyser readings, the values of m and C for the best straight line (called the regression of x on x_r) are determined as follows.

The deviation of a single point from the line is given by

$$x_r - (mx + C)$$

The sum of the squares of all the deviations is

$$\Sigma \left[(x_r - mx - C)^2 \right]$$

The best line corresponds to the minimum value of this summation.

Solution of the equations obtained by equating to zero the partial derivatives of the sum, with respect to m and C, leads to the expressions:

$$m = \frac{\Sigma(x)\, \Sigma(x_r) - n\, \Sigma(x\, x_r)}{[\Sigma(x)] - n\, \Sigma(x^2)}$$

$$C = \bar{x}_r - m\bar{x} = [\Sigma(x_r) - m\, \Sigma(x)]/n$$

A measure of the degree of correlation between x and x_r is given by the correlation coefficient, r.

$$r = \frac{\Sigma(x - \bar{x})(x_r - \bar{x}_r)}{\sqrt{[\Sigma(x - \bar{x})^2]\,[\Sigma(x_r - \bar{x}_r)^2]}}$$

$$= \frac{n\,\Sigma x\,.\,x_r - \Sigma x\,.\,\Sigma x_r}{\sqrt{[n\,\Sigma x^2 - (\Sigma x)^2]\,[n\,\Sigma x_r^2 - (\Sigma x_r)^2]}}$$

For perfect direct correlation, $r = 1$. To indicate significant direct correlation, r must be close to 1, since a false apparent correlation may arise between variables that are really truly independent. There are therefore critical minimum r values (depending on n) to indicate correlation [6].

To prepare the control chart, first the regression line is drawn, then two additional 'action' lines are drawn above and below it, placed so that 95% of all results should lie between them. Thus the action lines should be drawn parallel to the regression line and at distances of $\pm 2s$ vertically from the regression line (see Fig. 4.4); the value used for s should be that found in the test of the *repeatability* of the analyser (p. 40).

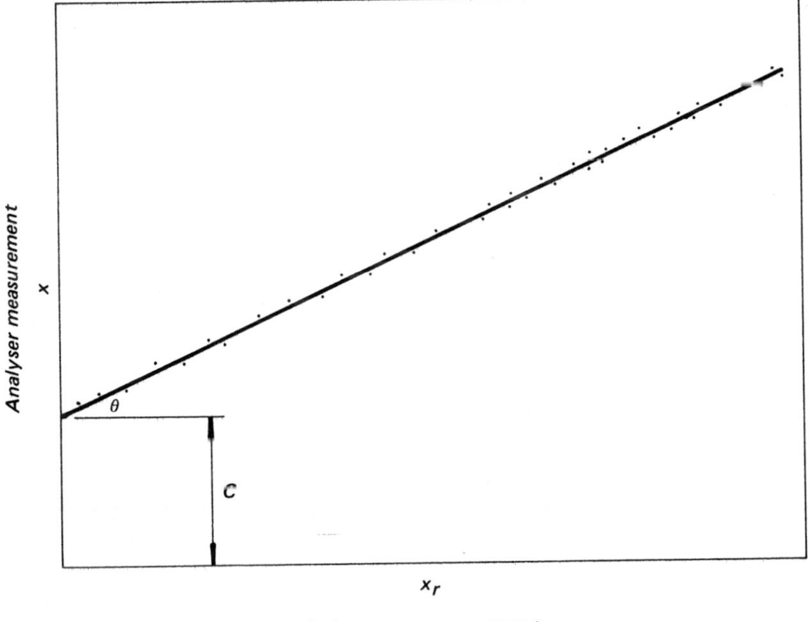

Fig. 4.3 – The regression of x on x_r. C is the *initial error* or *bias*.

4.4.3 Routine control checks

It is necessary to make a regular check of the function of the analyser. This may be done at intervals of between once per shift and once per week, depending on the reliability of the analyser. For each check, a sample is taken for laboratory analysis, and at the same time the analyser signal is noted. Both values are entered in the log, and also plotted on the control chart. If it is found that more than 5% of points fall outside the limits of the action lines, some corrective action (e.g. maintenance) must be taken.

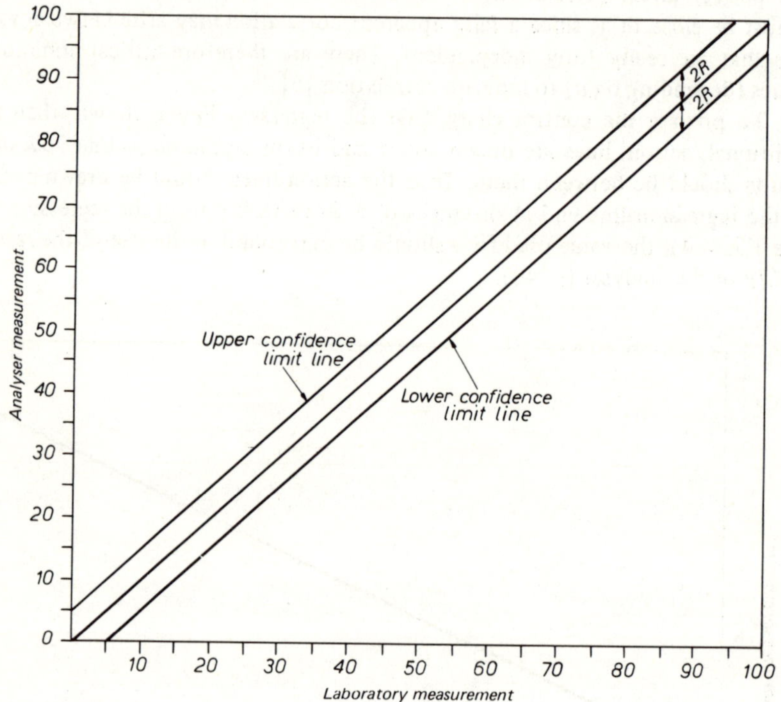

Fig. 4.4 – A control chart. *R* is the value for *s* determined in the measurement of repeatability.

4.4.4 Alternative forms of control chart

In some control charts, in addition to plotting 95% confidence lines (at $\pm\, 2s$), an extra pair of lines is plotted at $\pm\, 3s$, corresponding to 99.7% confidence. The $2s$ lines are then called 'warning lines' and the $3s$ lines, 'action lines'. Corrective action is taken if *one* point falls outside the action lines, or if two successive points fall outside the warning lines.

If it is important to have the earliest possible warning of a need for corrective action, a *cumulative sum*, or *cusum* chart, may be used. Such charts are discussed fully in [5].

REFERENCES

[1] J. Mandel, *Mater. Res. Stand.*, 1971, **11**, (Aug.), 8, 52.

[2] J. Mandel, *Mater. Res. Stand.*, 1972, **12**, (Sept.), 8, 61.

[3] L. A. Currie, in I. M. Kolthoff and P. J. Elving, eds., *Treatise on Analytical Chemistry*, 2nd Ed., Part 1, Vol. 1, Wiley, New York, 1978, p. 93.

[4] J. Mandel, in I. M. Kolthoff and P. J. Elving, eds., *Treatise on Analytical Chemisty*, 2nd Ed., Part 1, Vol. 1, Wiley, New York, 1978, p.243.

[5] O. L. Davies and P. L. Goldsmith, *Statistical Methods in Research and Production*, 4th Ed., Longman, London, 1972.

[6] K. Eckschlager, *Errors, Measurement and Results in Chemical Analysis*, Van Nostrand Reinhold, London, 1969.

[7] C. Liteanu and I. Rîcǎ, *Statistical Theory and Methodology of Trace Analysis*, Horwood, Chichester, 1980.

[8] D. L. Massart, A. Dijkstra and L. Kaufman, *Evaluation and Optimization of Laboratory Methods and Analytical Procedures*, Elsevier, Amsterdam, 1978.

[9] K. Echschlager and V. Štěpánek, *Information Theory as Applied to Chemical Analysis*, Wiley, New York, 1979.

[10] *Anal. Proc.*, 1980, 17, 166-200 (a series of short articles on analytical quality control).

[11] J. R. DeVoe, *Validation of the Measurement Process*, ACS Symposium Series, No. 63, ACS, Washington, D.C., 1977.

CHAPTER 5

Response time

If an analyser response is characterized by a single time constant (τ), then the growth of the response is described by the equation

$$x_t = x_0 (1 - e^{-t/\tau})$$

where x_0 is the input ($\equiv$ full response) and x_t is the response at time t.

When $t = \tau$, the response is $x_0 (1 - 1/e)$, which is $0.632\ x_0$, i.e. the response is 63.2% complete. Similarly, the response is 86.4% complete at $t = 2\tau$, and 95% complete at $t = 3\tau$. Likewise, the response is 50% complete when $e^{-t/\tau} = 0.5$, i.e. $t/\tau = \ln 2 = 0.693$, and this occurs at a time called t_{50} or $t_{\frac{1}{2}}$. Similar calculations can be made for the time taken to reach or traverse between other fractional responses, e.g. $t_{90} = 2.3\ \tau$, $t_{10-90} = 2.2\ \tau$, $t_{99} = 5\ \tau$. Because the growth of the response is exponential, it is advantageous in terms of time to use a response that is large but not complete. Thus if the absolute reading error is constant over the range used, use of the t_{90} response results in a relative error only 10% larger than that for 100% response.

In practice, the analyser response function may contain a pure lag time (dead time) and more than one time constant. The lag is the time delay between insertion of the input to a system and the appearance of the resulting output. It is often due to transmission lines, such as sample lines. The lag time (t_L) in a sample line obviously depends on the volume of the line and on the rate of sample transport through the system:

$$t_L \text{ (sec)} = \frac{\text{volume (m}^3)}{\text{volume flow-rate (m}^3/\text{sec})} \equiv \frac{\text{line length (m)}}{\text{linear velocity (m/sec)}}$$

Some manufacturers quote a short t_{50} response time but do not disclose a very long and important overall t_{90} response time, and care should be taken to check thoroughly. Sensor response can be rapid but cell response may be very slow (perhaps because of slow diffusion from the conditioning system to the sensor itself).

The overall response time of an analyser system will be dependent on the individual time constants of the component parts, and on the design of the system. It is most conveniently dealt with by means of **transfer functions**, based on the use of the Laplace transformation [1]. The transfer function $G(s)$, is defined as the ratio of the Laplace transforms of the output $[C(s)]$ and input $[M(s)]$.

$$G(s) = \frac{\text{output}}{\text{input}} = \frac{C(s)}{M(s)}$$

$G(s)$ represents the dynamic response of the system (the transient response) to changes in the input.

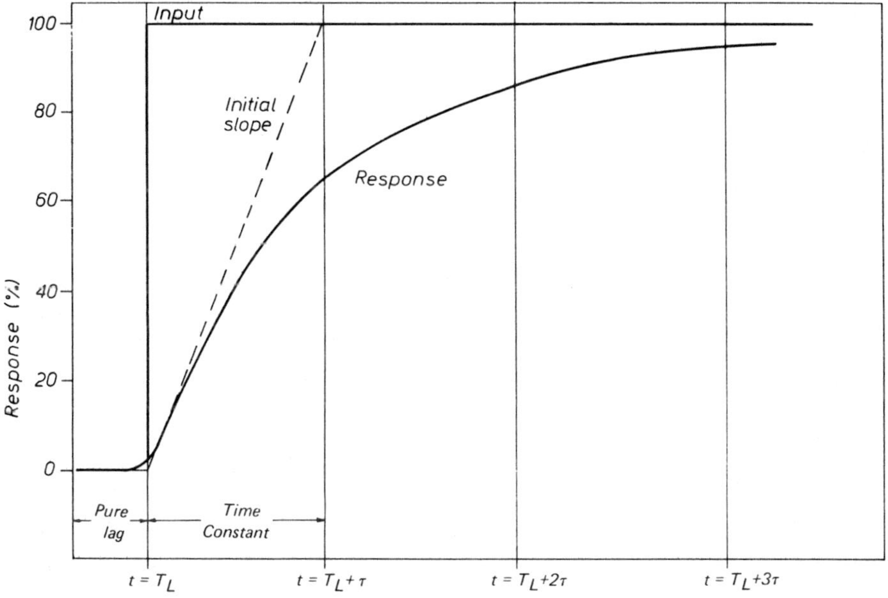

Fig. 5.1 — A response curve.

Many systems can be described by combinations of transfer functions, these being

proportional element	K
capacitance element	$1/Ts$
first order element	$1/(Ts + 1)$
second order element	$1/(T_1s + 1)(T_2s + 1)$
dead-time element	e^{-Ls}

where T is a time constant, L the dead time, and s the Laplace transform variable.

For systems arranged with process elements in series, the composite transfer function is the product of the individual transfer functions, and for elements in parallel it is the sum of the transfer functions.

Obviously if a system consists of a series of first-order lags, the overall response becomes slower. Hence

$$G(s) = 1/(T_1 s + 1)(T_2 s + 1) \ldots (T_n s + 1)$$

If there is a dead-time lag, the function is

$$G(s) = e^{-Ls}/(T_1 s + 1) \ldots (T_n s + 1)$$

This is exemplified by some commercial air-quality monitors [2] where the relation

$$G(s) = e^{-Ls}/(T_1 s + 1)(T_2 s + 1)$$

was found.

The transform function is translated back into terms of an overall time constant by using the tables for the inverse of the Laplace transform. A short table of transforms is given in Doebelin's book on system dynamics [3].

Dead times due to the sample line vary between say 0.1 and 3 minutes, depending on line length and flow velocity. Other dead times in analyser systems are usually shorter than that due to the sample line, but dead times resulting from retention times in a gas chromatograph can be as long as 30 minutes. Time constants range from ~ 1 sec for optical analysers with small flow-through cells to a minute or longer for some electrochemical analysers.

Methods for determining the time constants experimentally have been described. In step tests, measurement is made of the response changes resulting from rapid shift from one steady-state input to another. In pulse tests, the input signal is disturbed for a short period of time and then returned to its original value, and the output signal is monitored; the pulse must not cause saturation of any component of the monitor system, so that the output signal still displays transient response characteristics. The step-test data are processed by means of a semi-log plot of percentage of total response not reached, *vs.* time [4]. The pulse tests are analysed by a Fourier analysis of the frequency content of the input and output pulses [5, 6].

The overall response time for an analyser system is the sum of the individual component response times, and may typically be:

$$t_A = t_s \text{ for the sample conditioning system } (\sim 2 \text{ min})$$

$$+ t_c \text{ for the analyser cell } (\sim 1 \text{ min})$$

$$+ t_e \text{ for the electronics } (\sim 1 \text{ sec}).$$

For most purposes, chemical process systems can be described in terms of a dead-time plus first or second order term. A good account is given by Shinskey [7]

REFERENCES

[1]   R. H. Perry and C. H. Chilton, *Chemical Engineers' Handbook,* 5th Ed., McGraw-Hill, New York, 1973, p. 22–6 ff.

[2]   K. B. Schnelle, Jr. and R. D. Neeley, *J. Air Pollut. Contr. Assoc.,* 1972, **22,** 55.

[3]   E. O. Doebelin, *System Dynamics: Modeling and Response,* Merrill, Columbus, Ohio, 1973, p. 477.

[4]   N. A. Anderson, *Instr. Control Systems,* 1963, **36,** (November), 130.

[5]   W. C. Clements, Jr. and K. B. Schnelle, Jr., *I. & E.C. Proc., Design Devel.,* 1963, **2,** 94.

[6]   D. R. Coughanour and L. B. Koppel, *Process Systems Analysis and Control,* McGraw-Hill, New York, 1965.

[7]   F. G. Shinskey, *Process-Control Systems,* 2nd Ed., McGraw-Hill, New York, 1979.

CHAPTER 6

The standardization or calibration of gas analysers

The accuracy of an analyser depends on the accuracy of its calibration. In order to standardize a gas analyser and check its operation, various gas mixtures are required. The requirements vary according to the particular type of analyser and the nature of the test.

6.1 SEPARATION GAS

The separation gas is a gas mixture containing all the impurities expected. The mixture is required for checking chromatographic separations and to check the sensitivity of continuous analysers, such as infrared analysers, to the components expected to be present.

6.2 CALIBRATION OR SPAN GAS

Calibration or span gas is an accurately made-up mixture containing only the components to be measured, but many contain an inert diluent (i.e. giving zero response). The concentrations of the components have to be measured more precisely than is required for calibration of the analyser (Table 6.1). Calibration gas is needed to check the range of the analyser, so the concentrations of the components should be about 80% of the maximum values.

Table 6.1

Typical preparation tolerances and certification accuracies

Range	Typical preparation tolerance	Typical certification accuracy
10–50 ppm	± 20%	± 5%
50 ppm–50%	± 10%	± 2%
10–50%	± 5%	± 2%

Improved certification accuracy is usually available for a slight increase in price. Large cylinders are often more economical to use than small cylinders, because the cost of the cylinder and gas is mainly due to the cost of the analysis of the contents.

If the relative response of the analyser is known sufficiently accurately for all the components of interest in the sample, the calibration gas need consist of only a single component in the diluent. For a given required accuracy, the cost of the calibration gas will then be less.

6.3 CHECK GAS

This is a typical sample taken from a similar plant to check the performance of the complete analyser assembly before it starts to take samples from the plant. It may also be used to check the analyser during plant operation.

6.4 ZERO GAS

Zero gas is often the span gas diluent and contains none of the impurities expected. It is used to check the analyser zero. Nitrogen of suitable purity is often used as zero gas.

6.5 PRECISION GAS MIXTURES

Precision gas mixtures can be made up by the following methods.

(i) The partial pressure method.
(ii) The partial volume methods: by injection from a syringe; by using a calibrated pump or by using a flowmeter such as a variable-area flowmeter.

The temperature of the gases during the procedure should remain as constant as required for the final make-up accuracy.

The partial pressures of the various components and of the diluent gas must be carefully predetermined to ensure that the gases do not condense and that the compressibility factors are sufficiently near to unity.

6.5.1 Partial-pressure method

The concentrations of the various components in a gas mixture are proportional to the partial pressures of these components (Dalton's law). To make up a precise gas mixture by the partial-pressure method, the following procedure may be adopted.

(a) Select a suitable cylinder.
(b) Evacuate the cylinder to a suitably low pressure, so that the partial pressures of the remaining components are negligible. Sometimes, it is necessary to dilute the residual gases with a suitable diluent and to re-evacuate the cylinder to obtain a sufficiently low pressure.

(*c*) Admit the first gas component into the cylinder until the total gas pressure in the cylinder has increased by the partial pressure required for that component.

(*d*) Repeat stage (*c*) with each component in turn, until all the components have been put into the cylinder.

(*e*) Fill the cylinder to the required final pressure with a diluent gas.

(*f*) Allow the gas mixture to become sufficiently homogeneous before analysing it with a laboratory analyser of sufficient precision.

6.5.2 Partial-volume method by injection from a syringe

(*a*) and (*b*). As for the partial-pressure method just described.

(*c*) Fill the cylinder with diluent gas to atmospheric pressure.

(*d*) Inject the required gas component into the cylinder.

(*e*) Repeat stage (*d*) until all the components have been injected into the cylinder.

(*f*) and (*g*). As (*e*) and (*f*) for the partial-pressure method.

This method is only suitable for very dilute gas mixtures. For mixtures with higher concentrations of the components, it is possible to inject the components in the liquid phase, provided they vaporize completely on dilution.

6.5.3 Partial-volume method using a calibrated pump

The partial-volume method using a calibrated pump has been well developed by Wösthoff GmbH of Germany, who have published details of the method and of the equipment required [1, 2].

6.5.4 Partial-volume method using a flowmeter

(*a*) and (*b*). As in Section 6.5.1.

(*c*) Fill the cylinder with the various components and the diluent until the desired filling pressure has been obtained. The flow-rate of each gas is carefully monitored and controlled to ensure that the final mixture is as accurate as required.

(*d*) As for (*f*) in Section 6.5.1.

This method is only as accurate as the accuracy with which the individual flow-rates are measured and controlled.

6.6 CALIBRATION FREQUENCY

The frequency of calibration or standardization of an analyser depends on the stability of the analyser and on the accuracy required. It is usual to reduce the frequency of calibration after the analyser has settled down and the precision has been proved to be within the required limits.

For an infrared analyser and a gas chromatograph, the standardization intervals given in Table 6.2 are typical.

Table 6.2

Standardization periods for two types of analyser

Analyser	Accuracy %	Standardization interval		
		Week 1	Weeks 2–5	Thereafter
Infrared	± 0·1	4 hr	8 hr	24 hr
	± 0·5	8 hr	24 hr	7 days
	± 1	24 hr	24 hr	7 days
	± 2	24 hr	7 days	7 days
Gas chromatograph	–	3 days	7 days	28 days

6.7 CYLINDER LIFE

The life of a cylinder of gas (t days) is given by the equation:

$$t = \frac{V_C(P_F - P_D)}{P_D(Q_c + Q_1)}$$

where V_C is the cylinder volume in litres, P_F the maximum filling pressure in bars, P_D the regular delivery pressure in bar abs, Q_C the flow-rate of gas (litres/day) to the analyser, Q_1 the average flow-rate of gas leaking from the cylinder (litres/day) at the delivery pressure. Two examples are given below.

$$
\begin{array}{lll lll}
V_C & = & 48 & (l.) & 17 & (l.) \\
P_F & = & 150 & (\text{bar abs}) & 30 & (\text{bar abs}) \\
P_D & = & 5 & (\text{bar abs}) & 3 & (\text{bar abs}) \\
Q_C & = & 5 & (l./\text{day}) & 0.1 & (l./\text{day}) \\
Q_1 & = & 15 & (l./\text{day}) & 1 & (l./\text{day})
\end{array}
$$

$$t = \frac{48(150 - 5)}{5 \times 20} = 70 \text{ days} \qquad \frac{17(30 - 3)}{3(0.1 + 1.0)} = 139 \text{ days} .$$

It is often necessary to replace the cylinder before the cylinder pressure has dropped to the delivery pressure.

When the leakage rate is large and inaccurately known, the calculated lifetimes are useful mainly for detecting such leaks, especially at higher pressures. The leakage rate depends very much on the quality of the seals between the cylinder and the regulator; it can be reduced considerably by wrapping a single layer of thin Teflon tape around the threads of the high-pressure connections.

Cylinder life also depends on the constancy of the composition of the gas mixture. If there is a gas leak, small molecules will leak preferentially. Also unsaturated hydrocarbons are adsorbed onto the inside surfaces of cylinders. As a result of these effects, but also depending on the accuracy required, cylinder life is normally limited to about 150–300 days.

Hastings–Raydist Inc. make a wide range of calibrated gas leaks which are available for flow-rates from 10^{-4} to 10^{-9} ml/sec with an error of less than 10% and life of typically 0·5–2 years before the error increases by 10%.

6.8 STORAGE OF CALIBRATION LIQUIDS

When mixtures of liquids are stored in cylinders these should be only partially filled with the liquid. The remainder of the cylinder volume should contain an inert gas such as nitrogen at a pressure such that the liquid does not vaporize and thus fractionate (Fig. 6.1).

Sometimes it is better to separate the gas and liquid by a flexible diaphragm of a suitable material. For this purpose a Flamco Airfix expansion vessel may sometimes be used (Fig. 6.2).

The liquid volume is equal to the vessel volume $\times\ [(P_w - P_0)/P_w]$ where P_w = absolute working pressure and P_0 = absolute gas pressure with zero liquid volume.

If the vessel is pressurized with gas and then filled with liquid, the liquid will always be under pressure and the pressure will be a measure of the liquid volume.

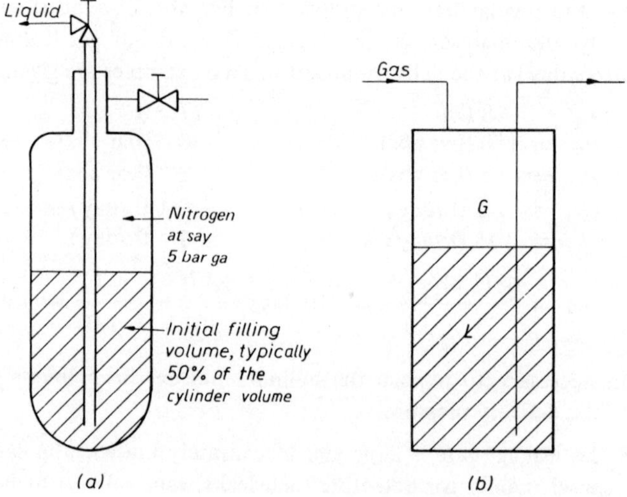

Fig. 6.1 – Methods of storing calibration liquids in cylinders.

6.9 USE OF LIQUID SAMPLE STORAGE TANK

For the calibration of liquid analysers, it is useful to provide a raised tank, near the analyser, which can be filled with sample to be analysed by the laboratory and by the process analyser (see Fig. 6.3).

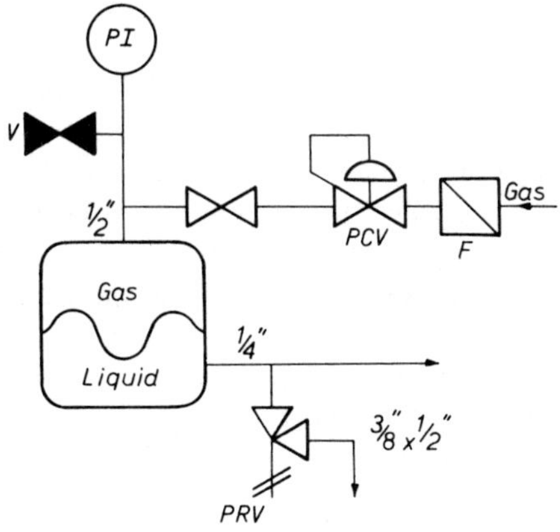

Fig. 6.2 – Use of the Flamco Airfix expansion vessel.

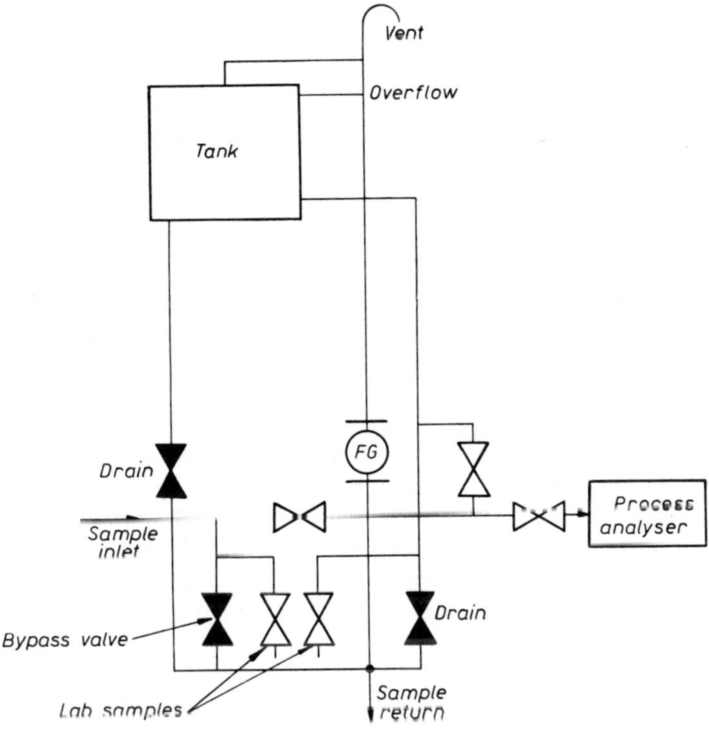

Fig. 6.3 – Use of an elevated tank close to the analyser.

6.10 GAS GENERATORS AND DIFFUSION TUBES

Hartmann and Braun have developed a proof-gas generator especially for the automatic calibration of air-quality analysers. The component of interest, which is dissolved in a liquid maintained at a constant temperature of $45°C$, slowly diffuses through a silicone-rubber membrane. It is then mixed with diluent and fed to the analyser. The instrument has been designed to generate calibration gas with a component concentration of 0, 50 and 100% of the analyser range. Each may be selected directly on the instrument or by remote control.

It is difficult to store a low concentration of H_2S in a diluent in a steel cylinder without rapid loss of the H_2S, even if the cylinder is properly lined. Thus this proof-gas generator should be very useful for the generation of dilute H_2S mixtures.

The Bendix Series 8850 permeation system uses Metronics Dynacal permeation tubes in a constant-temperature oven. (Fig. 6.4).

The dilution ratio is given by $F_p/(F_p + F_d)$, where F_p is the flow-rate of the permeation air and F_d is the flow-rate of the dilution air. Table 6.3 gives typical permeation rates at different temperatures.

Table 6.3

Permeation rates in ng/min per mm of tube length

	15°C	22°C	30°C
H_2S	11.5	21.0	45.7
SO_2	12.8	22.3	42.2

For H_2S it is necessary to use nitrogen instead of air as diluent. Permeation tubes are also available for NO_2, C_3H_8, CH_3SH, Cl_2, NH_3, HF and propylene.

6.11 MOISTURE GENERATORS

DuPont make a moisture generator for use in the calibration of analysers, at low concentrations of moisture. It is illustrated in Fig. 6.5.

The diffusion cell has four membranes which permit controlled diffusion of the moisture from the bowl into the gas stream. Each membrane is separated from the adjacent one by two aperture plates each 1/16 in. thick, with a central hole, the diameter of which determines the rate at which the moisture diffuses into the gas stream.

The diffusion cell and its bowl and the heat exchanger are in a water-bath, the temperature of which determines the moisture concentration in the gas, which is a linear function of temperature. There are various models of diffusion cell, covering different calibration ranges (which overlap to cover the range from about 1 to 1000 ppm v/v).

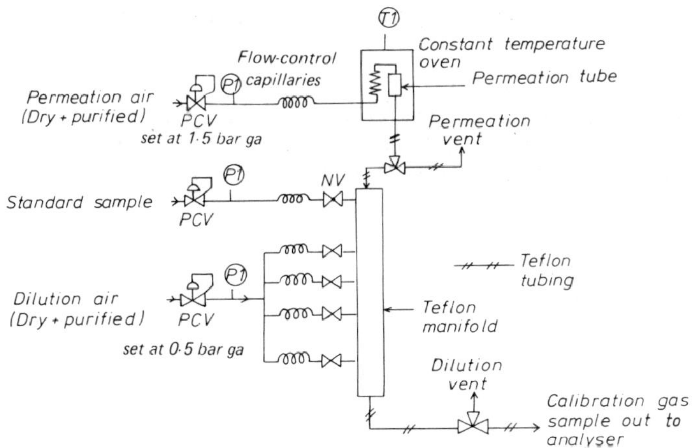

Fig. 6.4 — Bendix Series 8850 permeation system.

The British Oxygen Company supplies stable low-concentration mixtures of water in fairly pure nitrogen in aluminium cylinders. These and mixtures of reactive gases are available in preconditioned and specially lined "Spectra-Seal" cylinders, which were developed as a result of co-operation between Airco (part of BOC International) and the U.S. Bureau of Standards.

Wechter and Kramer [3] have shown that, provided none of the components in the cylinder is allowed to condense, the long-term stability over 2 years is about ± 2 μl/l. at the 50–μl/l. level, if the storage temperature is reasonably constant (e.g. in the range 10–30°C).

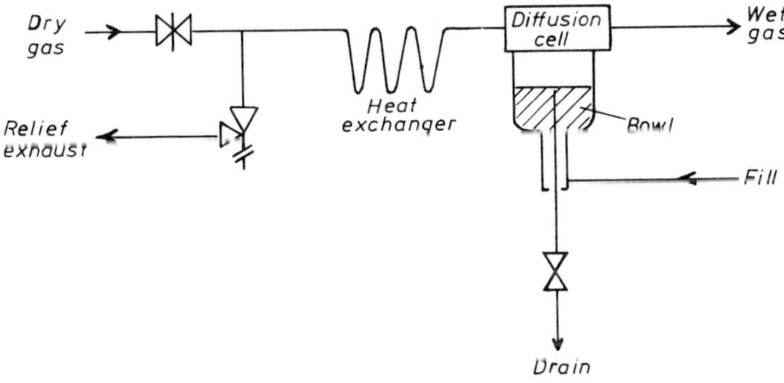

Fig. 6.5 — DuPont moisture generator.

REFERENCES

[1]   H. Wösthoff OHG (D–4630 Bochum 1, GFR), *Catalogue 511e* (DIGIMIX gas-mixing pumps).
[2]   H. Wösthoff OHG, *Leaflet 511–2.03*.
[3]   S. G. Wechter and F. Kramer, Evaluation of gas phase moisture standards prepared and treated aluminium cylinders, 21st Nat. Symp. Anal. Instr. Divn. ISA, King of Prussia, 8 May, 1975.

CHAPTER 7

Gas purification

Gases for use in analysers usually require some degree of purification before use. Moisture is the most important impurity to be removed; others include oxygen, carbon dioxide, carbon monoxide, etc.

A molecular sieve such as MS 5A is one of the best moisture absorbents. Special drying packs, supplied by Matheson and others, contain such a molecular sieve, along with some activated carbon to adsorb unsaturated hydrocarbons. Some suitably dyed molecular sieve or silica gel is added to act as an indicator of exhaustion of the drier. However, it should be realized that cobalt salts are not very efficient indicators. The life of the drier is typically from 1 to 6 months with a gas flow of 2·5–30 l./hr, depending on the moisture content of the gas supply and the size of the drier. A typical unit is shown in Fig. 7.1.

Molecular sieves can be regenerated in an oven. Every 2–3 years, according to the extent of hydrocarbon leakage, the molecular sieve should be washed with water and dried in an oven. Silica gel can be regenerated or dried in an oven but it is better to replace it altogether. Activated carbon should be replaced when contaminated.

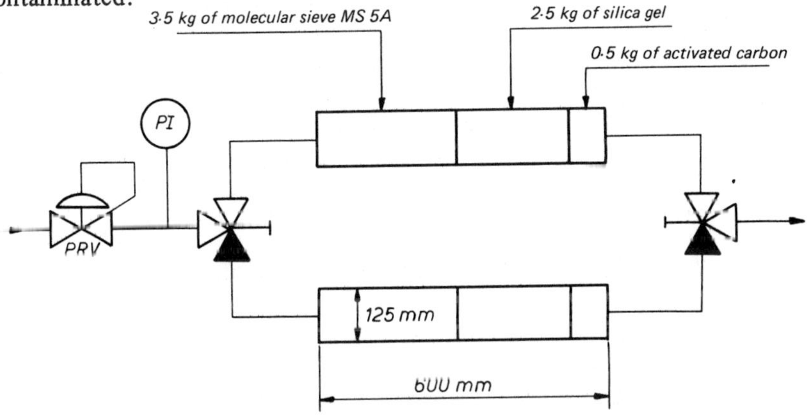

Fig. 7.1 – A typical gas drier.

A 1:1 mixture of phosphorus pentoxide with a coarse firebrick has been used successfully for drying gases. The life of the drier may be from 3 weeks to 3 months, depending on the gas flow-rate, which is typically 1·0–45 l./hr, and also depending on the moisture content of the gas supply. The white phosphorus pentoxide gradually becomes brown as it absorbs water. Anhydrous magnesium perchlorate [$Mg(ClO_4)_2$] is also used for drying gases; it is effectively exhausted when it reaches the composition of the hexahydrate, i.e. it will absorb $\sim$ 480 g of water per kg. To prevent risk of explosion in the regeneration of spent magnesium perchlorate that has been exposed to organic vapours, the method given by G. F. Smith (*Talanta,* 1960, **5**, 189) should be used.

Drierite is a commercial form of anhydrous calcium sulphate which remains crystalline even after having absorbed a large amount of water. A self-indicating Drierite changes from blue to pink when it is saturated with moisture. It may be regenerated by heating at 200°C.

Other possible drying agents include anhydrous sodium carbonate, copper sulphate, sodium sulphate and disodium hydrogen phosphate. A carbonate should not be used if carbon dioxide is to be measured.

A Deoxo hydrogen purifier may be used to remove traces of **oxygen** down to 1 ml/m^3. It converts free oxygen into water by the catalytic action of palladium at room temperature.

A good trace-oxygen scavenger is manganese(II) oxide (Fig. 7.2). Obviously, it must be kept isolated from air, otherwise the life of the scavenger will be very short. The manganese(II) oxide may be regenerated with hot pure hydrogen with the tube heated externally. The water formed must be carefully removed.

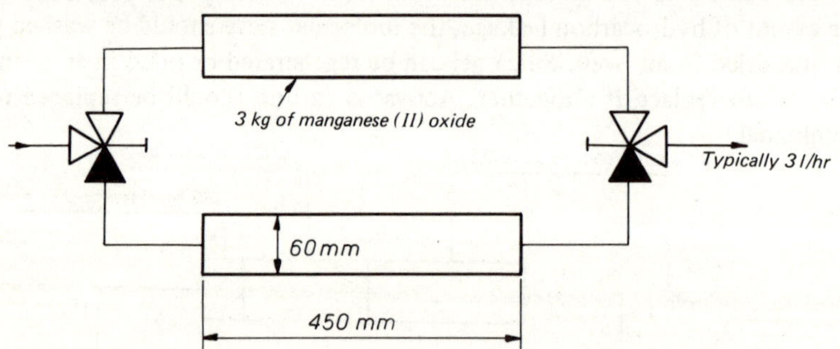

Fig. 7.2 – Use of manganese(II) oxide as an oxygen scavenger.

Sodium hydroxide is a popular absorber of **carbon dioxide**. It is often used on asbestos and is then described as soda asbestos or ascarite. To be effective, at least some of the sodium hydroxide must be moist. If water is also to be removed, it is essential to back the soda asbestos with an equal length of an efficient water absorber. Molecular sieves also absorb carbon dioxide.

Hopcalite will catalytically oxidize **carbon monoxide** to carbon dioxide. Hopcalite consists of 50% MnO_2 + 30% CuO + 15% Co_2O_3 + 5% Ag_2O. The exothermic reaction is described by

$$2\,CO + O_2 \longrightarrow 2\,CO_2 + 136\ \text{kcal}$$

This is the basis of the MSA carbon monoxide analyser, which measures the heat of the reaction. Iodine pentoxide is also used, at about $150°C$:

$$5\,CO + I_2O_5 \longrightarrow 5\,CO_2 + I_2$$

Hot copper combines with sulphur and thus can be used to remove **sulphides** from a gas stream:

$$H_2S + 2\,Cu \longrightarrow Cu_2S + H_2$$

$$CS_2 + 4\,Cu \longrightarrow 2\,Cu_2S + C$$

$$COS + 2\,Cu \longrightarrow Cu_2S + CO$$

The copper can be regenerated by reduction with hydrogen.

Hot copper can be used to remove **oxygen** from an impure process stream down to about $0.25\ \text{ml/m}^3$,

$$2\,Cu + O_2 \longrightarrow 2\,CuO$$

It may be regenerated by passing over it a stream of 5–10% hydrogen in an inert gas such as nitrogen.

The cobalt oxide (Co_3O_4) filter (Fig. 7.3) for oxidizing hydrocarbons etc. is usually run at a temperature of $200\text{–}300°C$.

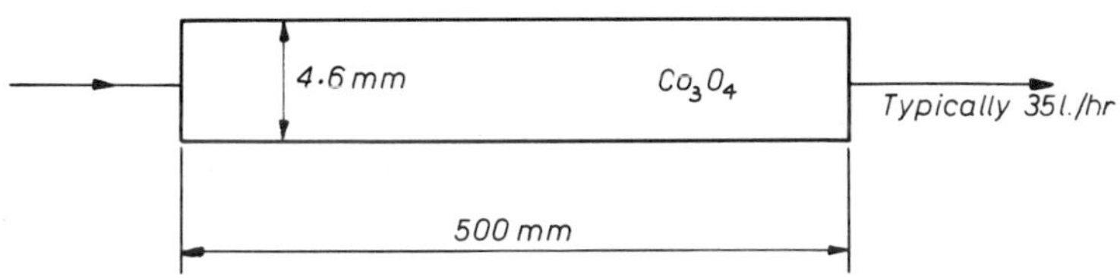

Fig. 7.3 – Cobalt oxide catalytic filter.

Potassium permanganate and potassium dichromate have been used as low-temperature oxidizers for organic compounds.

Helium may be purified by passing it through a quartz capillary tube (Fig. 7.4) heated to $300\text{–}600°C$. The helium diffuses through the wall of the hot capillary tube but the impurities (except hydrogen) do not because their molecules are too large. If hydrogen must also be removed, the supply gas should first flow through heated copper oxide so that most of the hydrogen is oxidized to water.

Pure **hydrogen** may be produced by the electrolysis of demineralized water. In the Metals Research Ltd. Gaspak-H hydrogen generator, the atomic hydrogen diffuses through the walls of a palladium-alloy tubular cathode. The hydrogen chamber is limited to a capacity of 600 ml and a pressure of 10 bar ga,

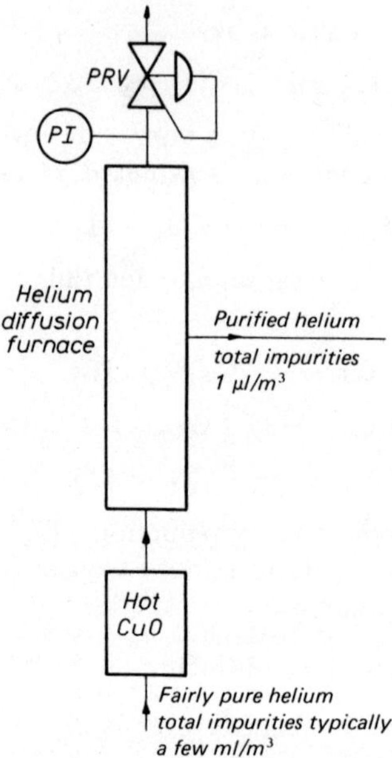

Fig. 7.4 — A helium purifier.

both of which are much smaller than the capacity and maximum pressure of normal hydrogen cylinders. The rate of flow of hydrogen delivered is 1·8–9 l./hr.

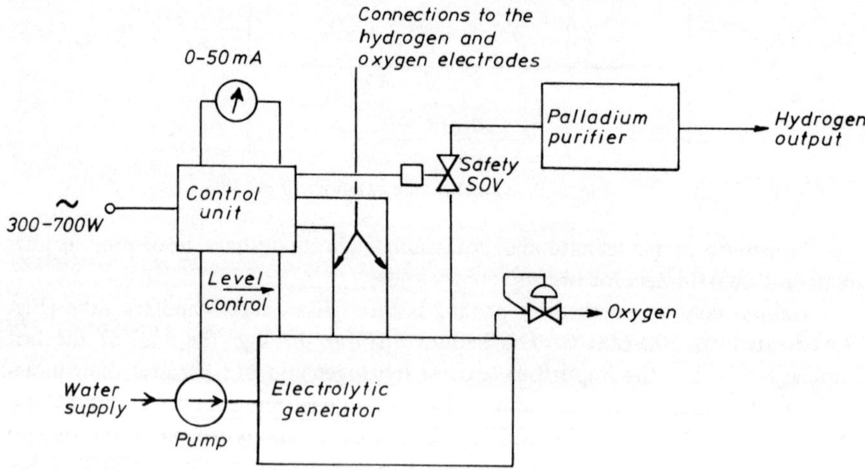

Fig. 7.5 — The Bendix hydrogen generator.

The Gaspak-H contains a demineralizer and thus can operate from normal tap water. It has a 2-litre reservoir which is sufficient for 5 days' continuous operation.

In the Bendix hydrogen generator (Fig. 7.5), the palladium purifier, which is an optional extra, consists of heated palladium tubing through which the hydrogen diffuses. The hydrogen produced is claimed to be 99·9999% pure. Water carry-over with the generated hydrogen is minimized by operating the generator at a pressure of about 14 bar ga.

Milton Roy make a hydrogen generator (Fig. 7.6), which is also sold by Matheson, in which the palladium-alloy diffusion tube also acts as a generator cathode. A pressure switch limits the maximum generator pressure and controls it to better than ± 0·5 bar. The pressure of the hydrogen delivered is held constant to about ± 0·002 bar.

Hartmann and Braun have also developed an electrolytic hydrogen generator. It has a capacity of 80 ml/min at 2 bar ga. The pressure controller regulates the current fed to a Varta fuel cell. The hydrogen is saturated with water at the cell operating temperature.

General Electric make a hydrogen generator (Fig. 7.7) which does not use a delicate palladium diffuser. The electrolyte is a solid polymer. The life of the electrolytic cell is considerably improved if it is supplied with demineralized

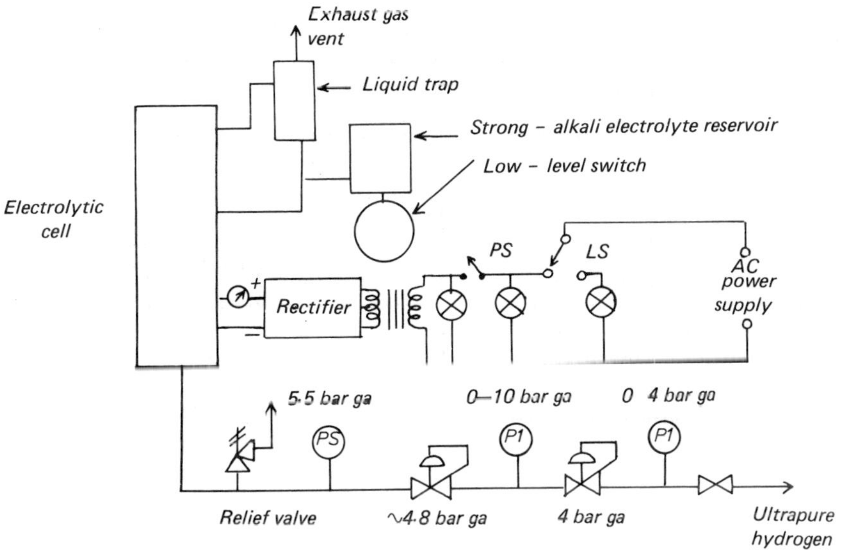

Fig. 7.6 — The Milton Roy hydrogen generator.

water. The demineralizer should preferably be provided with an integral purity
meter. The advantages of using a hydrogen generator instead of cylinders of
ultrapure hydrogen are:

(i) the hazards of storing hydrogen gas are reduced,
(ii) no cylinder contamination,
(iii) considerably lower cost, typically about one fortieth.

Hydrogen saturated with water may be acceptable for some flame-ionization
analyser applications but it is not suitable for use as a chromatographic carrier
gas, which must be dry.

Johnson Matthey Metals Ltd. make a range of diffusion units that use a
23% silver/77% palladium alloy membrane which does not distort as pure palladium membranes do when they are thermally cycled.

The use of an oxygen scavenger upstream of the purifier increases its life.
The use of an activated-charcoal purifier reduces hydrocarbon build-up.

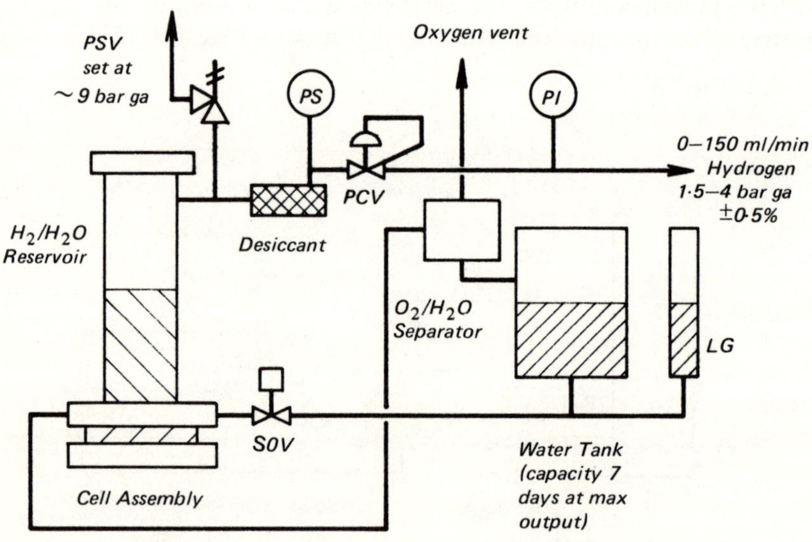

Fig. 7.7 – The General Electric hydrogen generator.

CHAPTER 8

Sample handling and conditioning

8.1 INTRODUCTION

A process analyser system normally has six main parts:

   (i) the sampling probe;
  (ii) the sample-transport line;
 (iii) the sample-conditioning system;
 (iv) the analyser itelf,
  (v) the analyser control unit or the programmer;
 (vi) the associated output equipment.

The sampling probe must ensure that the sample taken is truly representative of the process fluid. The sample-transport line must transport the sample from the take-off point to the analyser in an acceptable time and without affecting the composition of the sample appreciably.

The sample conditioning system is designed to ensure that the sample is acceptable to the analyser and is still truly representative. It removes solids and water as required and reduces or controls the pressure and temperature of the sample according to the analyser requirements. Sometimes the analyser is flow-sensitive, so the sample flow-rate must be controlled with suitable accuracy. The limited or controlled condition often has to be indicated. The downstream equipment often has to be protected against component failure, and sometimes a sample or component failure should be annunciated. The conditioning system should also ensure, where necessary, that the sample at the transport line inlet is a single-phase sample. The sample-conditioning system components are often located near the analyser, but sometimes near the sample take-off point instead.

The analyser then accepts the conditioned process sample and produces an appropriate output signal.

The simplest possible system is usually the most reliable, so care should be taken not to introduce unnecessary complexity.

8.2 SAMPLING PROBES

The sampling probe must acquire a sample that is truly representative of the process stream, and containing the minimum of undesired solids.

If the process stream is not homogeneous and the composition of the stream varies appreciably across the stream section, there should be several sampling apertures at different distances from the centre of the process pipe to ensure as true an average sample as possible. Sometimes it is necessary to take samples from various locations in the sample stream and then to mix them in suitable proportions according to the known distribution.

If the process fluid is multi-phase, an isokinetic sample of compositions in the stream should be taken, (i.e. the sample should flow into the sample line at the velocity of the process stream). The velocity of the sample at the sampling aperture should be equal to the velocity of the process fluid within about 20%.

In horizontal lines and in vessels, the various phases of a process stream tend to separate, especially at low velocities or as a result of centrifugal action. The gas phase rises to the top, and the solids and other denser phases, such as condensate in gas lines, tend to drop to the bottom.

Thus, gas samples should be taken from the top and liquid samples should be taken from the side of horizontal lines (Figs. 8.1 and 8.2). If the probe takes samples from across most of the process stream, then convenience is the determining factor for orientation.

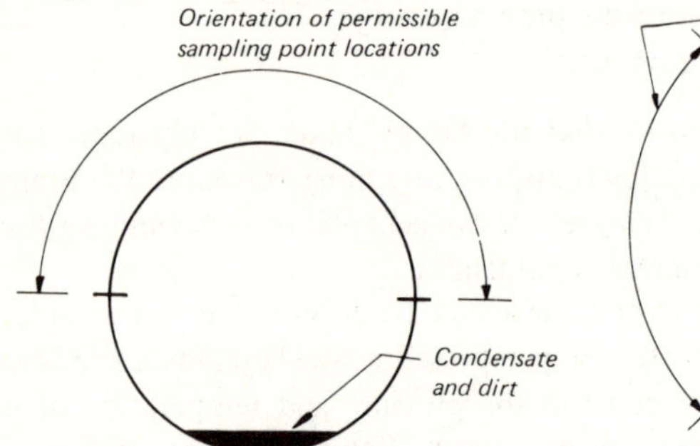

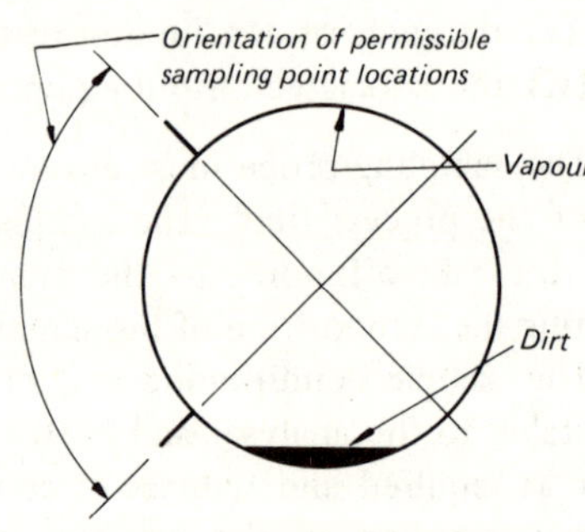

Fig. 8.1 – Sampling from a horizontal gas line.   Fig. 8.2 – Sampling from a horizontal liquid line.

It is essential that the sample taken from the process stream is up-to-date, true and clean. Thus, the sample should not be taken from the sides of the vessel or pipe where dirt could be picked up and where the sample would, quite probably, be non-representative owing to surface effects, temperature gradients, etc.

The entry holes of the probe should not face the process stream, otherwise they could get clogged with dirt. However, if the solid content is to be measured, the holes must face the stream.

Well designed and located probes do not normally require frequent cleaning. If they do, then a means must be devised of withdrawing the probe through a gate valve and stuffing box (Fig. 8.3).

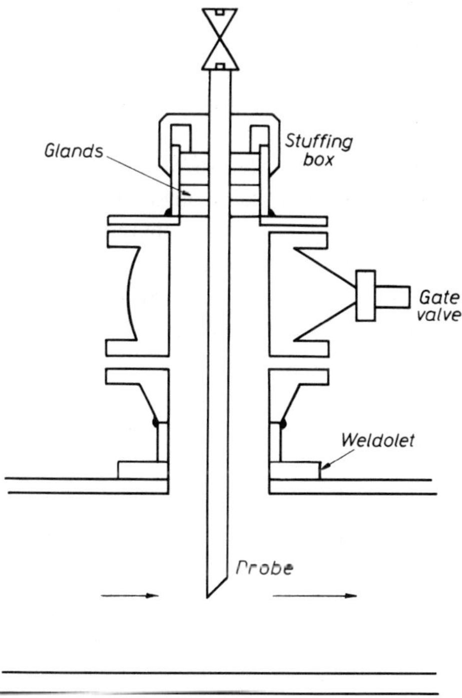

Fig. 8.3 – Gate value and stuffing box for withdrawing sample probe.

If the probe requires frequent cleaning and a withdrawable probe is un-suitable, e.g. in high-temperature sampling after cracking or coking, rodding facilities should be available (Fig. 8.4). The rod is inserted into the gland before the valve is opened so that there is a minimum amount of leakage at the rodding connection.

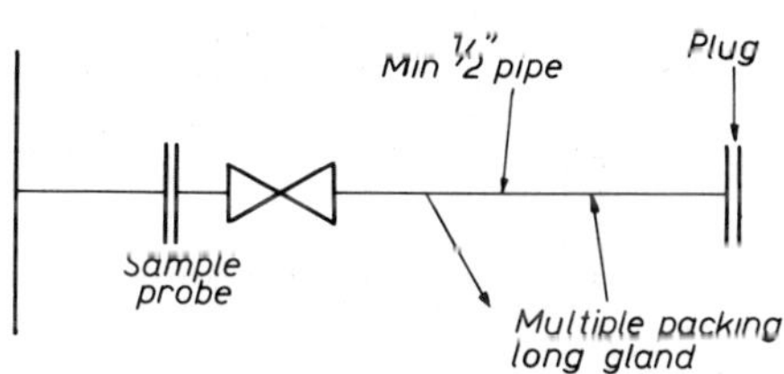

Fig. 8.4 – Rodding facility.

A line sampling probe (Fig. 8.5) should normally be fitted to the line by means of a flanged nozzle.

A sample take-off point without a probe, and most sample-return points, may consist of a ½ or ¾ in. gate-valve screwed on to a threaded nozzle and seal-welded at the process side.

For simple gas sampling, it is often sufficient to make the probe from some clean smooth pipe with the end cut at an angle of about 45° (Fig. 8.6). The size of pipe used depends on the amount of sample required, the acceptable lag, and the required rigidity. The probe must be strong enough not to break by vibration fatigue at the maximum fluid velocity, and also be strong enough to withstand normal field maltreatment; $\frac{1}{4}$ in. (6 mm) or $\frac{1}{2}$ in. (12 mm) nominal bore schedule 80 stainless-steel pipe is often used.

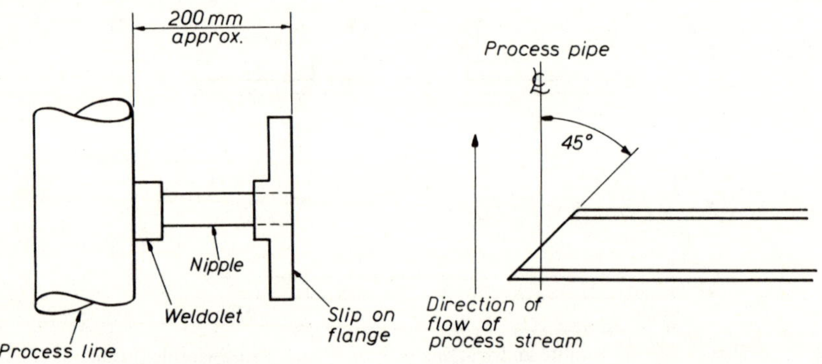

<table>
<tr><td>

Fig. 8.5 – Method of fitting sampling probe to line.

</td><td>

Fig. 8.6 – Typical sample probe (₵ = centre line of pipe).

</td></tr>
</table>

An alternative probe for vapour and liquid samples is shown in Fig. 8.7. The number of entry holes used would depend on the size of the process pipe. Holes of $\frac{1}{8}$ in. (3 mm) diameter should be suitable for $\frac{1}{4}$ in. probes and $\frac{1}{4}$ in. (6 mm) holes should be suitable for $\frac{1}{2}$ in. probes.

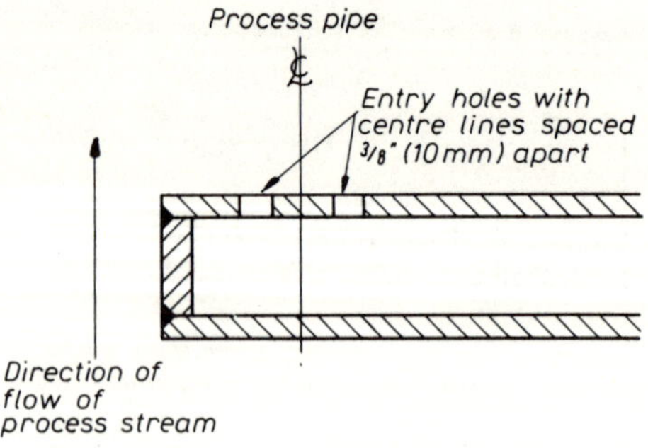

Fig. 8.7 – Alternative sample probe.

The preferred insertion length L (Fig. 8.8) and the maximum number of entry holes may be obtained from Table 8.1.

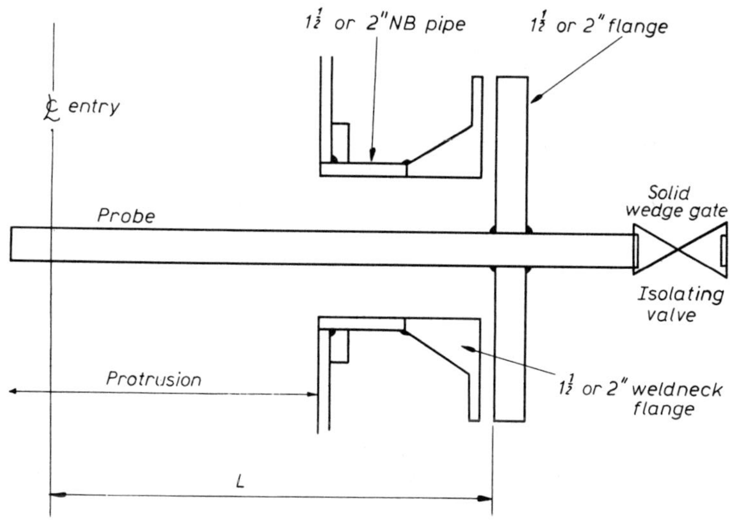

Fig. 8.8 – Method of fitting alternative probe. The probe holes lie between 0 and 0.3 line diameters from the centre line. (NB = nominal bore.)

Table 8.1

Probe insertion lengths and maximum numbers of entry holes

Process line size (inches)	2	3	4	5	6	8	10	12
Length L in.	$8\frac{1}{4}$	$8\frac{1}{4}$	$8\frac{1}{4}$	$10\frac{1}{4}$	$10\frac{1}{4}$	$10\frac{1}{4}$	14	14
mm	210	210	210	260	260	260	356	356
Number of entry holes	1	2	4	4	4	5	6	6

If sampling across most of the process stream is required then several holes should be provided, evenly spaced across the stream.

In large gas lines, such as flue ducts, it may be necessary to take several samples in order to obtain a sufficiently representative sample at the analyser. The flow-rates of the individual samples are adjusted according to the velocity and concentration of the process stream at the individual sampling points. The stream velocity profile is shown in Fig. 8.9, and a mixing system in Fig. 8.10.

The average concentration of the sample for the whole of the process stream is given by

$$\bar{c} = \frac{\Sigma Vc}{\Sigma V}$$

where V = velocity of the process stream at sampling location, c = concentration of the component in the process stream at the sampling location.

Similar precautions may have to be taken to make allowance for layering in tanks. It is normal practice to analyse tank samples in the laboratory and not automatically in the field.

When a moisture-analyser sample is to be taken, it is necessary that the take-off point is near to and downstream from a pump or valve or other source of turbulence so that the moisture concentration is uniform across the process line, with no migration towards the pipe walls owing to the polarity of the water.

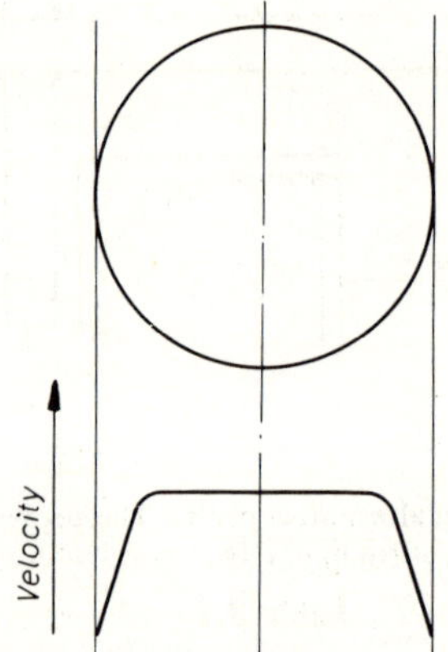

Fig. 8.9 – Variation of stream velocity across the process line.

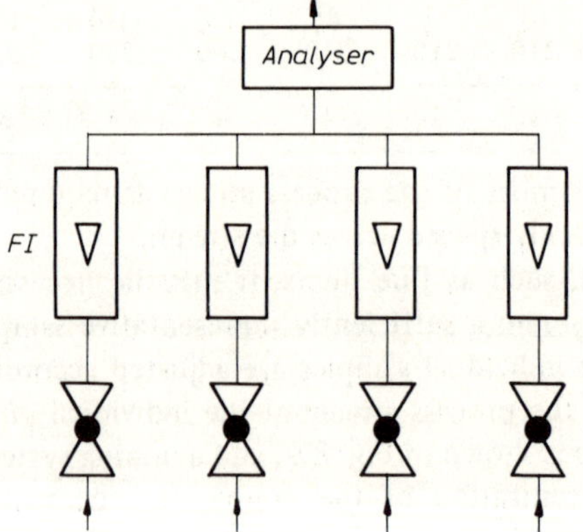

Fig. 8.10 – Mixing from several sample points.

8.3 SAMPLE TAKE-OFF

Only in relatively few cases can the analyser be mounted directly in the process line, because the sample flow-rate required is usually much smaller than the flow-rate of the process fluid. Also, the analyser often cannot work at the process temperature and pressure. The presence of dirt, etc., can also prevent the use of such an in-line analyser.

The analyser flow-stream must therefore generally be mounted in parallel to the process stream, and some means of pressure (or flow-rate) reduction must be incorporated.

The sampling system may by-pass a control valve (Fig. 8.11) but the pressure drop is often too small and too variable to be acceptable. The minimum pressure drop is typically approximately 0.5 bar and the average pressure drop is approximately 1 bar. The resultant small leakage resulting from by-passing the valve must be acceptable to the process control unit.

The process fluid is mixed in a line after a valve or other restriction, especially when there is a large differential pressure. If the differential pressure is large, a gas can cool considerably and a liquid can flash.

A restriction orifice can be used to produce the necessary pressure drop (Fig. 8.12), but this could result in a need for a process pump or compressor of increased size and cost. For this reason it might be better to use a sample pump (Fig. 8.13, a, b).

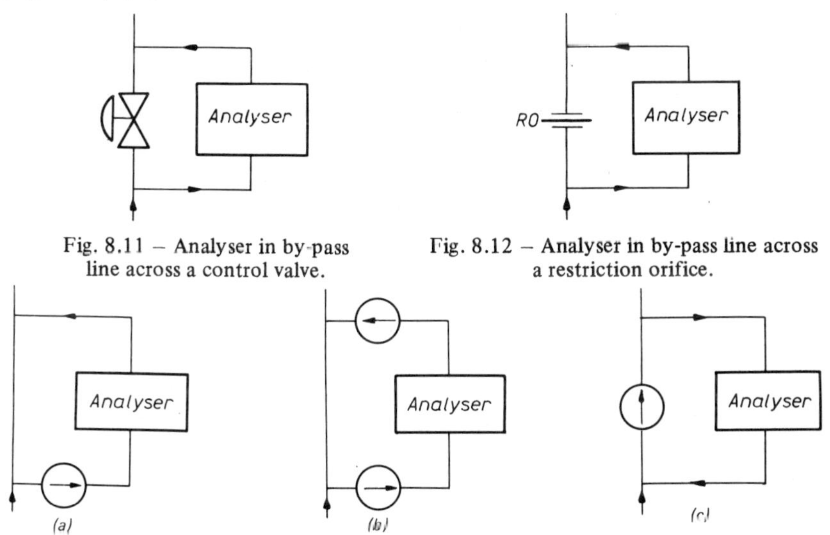

Fig. 8.11 – Analyser in by-pass line across a control valve.

Fig. 8.12 – Analyser in by-pass line across a restriction orifice.

Fig. 8.13 (a, b) – Two methods of using a sample pump. (c) Use of a process pump for sampling.

An alternative is to use part of the capacity of a process pump (Fig. 8.13, c). When the liquid sample is returned to the suction side of a pump, however, there must be no appreciable vapour present in it and it must not vaporize appreciably. If the returned sample contains vapour, it should be returned to a vapour dis engaging drum, such as an accumulator, or to a stripper (Fig. 8.14).

If a sample is taken from across a compressor, the best take-off point is often after the discharge knock-out drum where the sample has been cooled and had the free water removed from it (Fig. 8.15). The return should always be at the inlet of the suction knock-out drum.

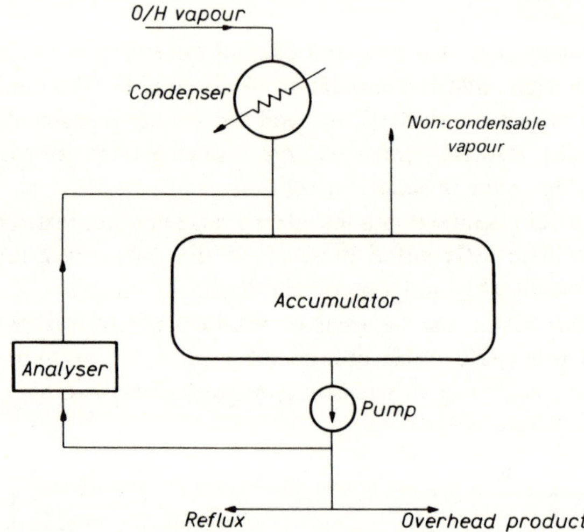

Fig. 8.14 — Returning a liquid sample containing some vapour to an accumulator rather than to the pump inlet.

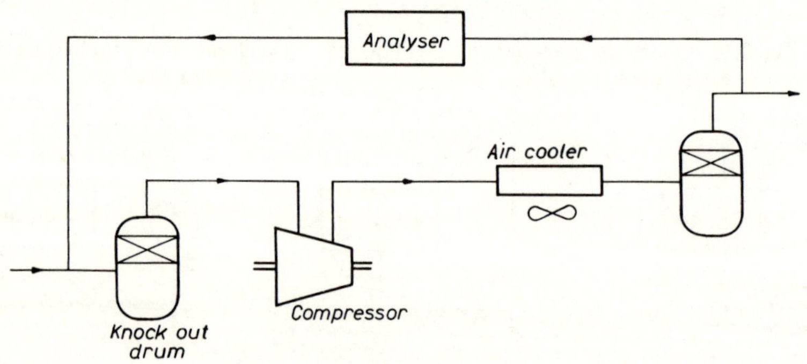

Fig. 8.15 — Returning a vapour sample containing some liquid to a knock-out drum rather than to the compressor inlet.

8.4 SAMPLE LINES

A sample line is used to transfer the sample from the take-off point to the analyser. If the sample fluid is liquid and can flash, or if the sample contains some vapour phase, then flow should be in an upward direction to allow free flow of the fluid without a vapour lock, but if the sample fluid contains solid particles, the flow should be in a downwards direction so that the particles are swept out of the system before they can accumulate and coagulate, thereby blocking the system.

If these flow directions cannot be maintained, then precautions must be taken; for instance, flushing connections could be added. If precipitation or polymerization takes place, frequent flushing is necessary. Flushing should normally be in the reverse direction, away from the smaller orifices.

Tight shut-off vent valves should be provided at the highest points to vent accumulated gases, and tight shut-off drain valves should be provided at the lowest points to remove condensate and accumulated solids. Knock-out pots are often necessary. Vent and drain valves should be low-leakage ball valves. Dead volumes should be minimized and traps avoided.

For a particular sample-line length and a required sample transport time, the sample fluid velocity can be calculated. The size of the line may then be determined according to the allowable pressure drop along the line and also according to whether pipe or tubing is to be used. For vapour samples, it is generally possible to get an adequate sample velocity without an unacceptably high product loss. For liquid samples, the liquid flow-rate has to be more carefully determined.

Turbulent flow produces better stream mixing than laminar flow, but it is not always economic to take a large enough sample flow-rate for turbulent flow to exist. In fast sample loops, where the sample is returned to the process, especially when the minimum product flow-rate is much larger than the sample flow-rate, turbulent flow is often justified.

For most clean gases and vapours, $\frac{1}{4}''$ o.d. $\times$ 0.049$''$ wall thickness type 316L stainless steel tubing is suitable. The bore of the tubing must be smooth and clean, so bright annealed tubing should be specified. Table 8.2 lists typical flow-rates for various tubes and pipes

Table 8.3 gives similar information for liquid samples. The sample velocity may be increased to reduce the transport lag time provided that the increased head loss is acceptable.

Table 8.2

Typical sizes and flow rates of gas sample lines.

Tubing size o.d. (in.)	Pipe size Nom. bore (in.)	Typical vapour sample flow-rate (m³/hr)	Resultant dead-time (sec/m)
$\frac{1}{4}$		0.09	0.5
	$\frac{1}{4}$	0.18	1.0
$\frac{1}{8}$		0.24	1.2
	$\frac{1}{2}$	0.3	1.8

Table 8.3
Velocity, rate of flow and head loss for some pipe sizes.

Pipe size Nom. bore (in.)	Velocity (m/sec.)	Rate of flow (m^3/hr)	Head loss (m H_2O/100 m pipe)
$\frac{1}{2}$	0.6	0.3	8
$\frac{3}{4}$	0.9	0.9	12
1	1.2	2.0	14
$1\frac{1}{2}$	1.5	6.3	14
2	1.8	12.6	14

(Note: 10 m $H_2O \cong$ 1 bar).

A method used for determining line sizes and the resultant flow-rates is illustrated below:

$$\text{linear velocity (m/sec)} = \frac{\text{distance in m}}{\text{time lag in sec}}$$

This must be multiplied by 3600 to get the linear velocity in m/hr. For the selected tubing or pipe, the volume per unit length (m^3/m) is known (Table 8.4) and thus the volume flow-rate can be calculated.

From the density of the fluid the mass flow-rate may be calculated.

Mass flow-rate (kg/hr) = volume flow-rate (m^3/hr) × density (kg/m^3).

Example

    Required time lag = 60 sec
    Distance between sample point and analyser = 100 m
    Required velocity = $\dfrac{100}{60} \cong$ 1.7 m/sec
    Selected pipe = $\frac{1}{2}$ in. schedule 80
    Volume per unit length of pipe = $150 \times 10^{-6}\ m^3$/m
    Volume flow-rate = $6000 \times 150 \times 10^{-6} = 0.9\ m^3$/hr
    Density = 800 kg/m^3
    Mass flow-rate = 720 kg/hr

Table 8.4
Volume per unit length of tubing and pipe[†]

Tubing size (in.)	Pipe size[‡] (nominal bore, in.)	Outer diameter (mm)	Wall thickness (mm)	Bore (mm)	Volume per unit length (10^{-6} m³/m)
$\frac{1}{8}$			0.56	2.06	3.33
$\frac{1}{8}$			0.71	1.75	2.41
$\frac{1}{4}$			0.81	4.72	17.5
$\frac{1}{4}$			1.22	3.91	12.0
$\frac{3}{8}$			0.91	7.70	46.6
$\frac{3}{8}$			1.22	7.09	39.5
$\frac{3}{8}$			1.63	6.27	30.9
$\frac{1}{2}$			1.22	10.26	82.7
$\frac{1}{2}$			1.63	9.44	70.0
$\frac{1}{2}$			2.03	8.64	58.6
1			3.18	19.05	285
	$\frac{1}{4}$	13.7	3.02	7.7	46.6
	$\frac{3}{8}$	17.1	3.20	10.7	89.9
	$\frac{1}{2}$	21.3	3.73	13.8	150
	$\frac{3}{4}$	26.7	3.91	18.9	281
	1	33.4	4.55	24.3	464
	$1\frac{1}{2}$	48.3	5.08	38.1	1140
	2	60.3	5.54	49.2	1901

†Volume flow-rate (m³/hr) = velocity (m/hr) × volume per unit length (m³/m).
‡Schedule 80 pipe.

As already pointed out, for a true sample to be obtained, the flow should be turbulent to ensure adequate mixing. Though this is readily achieved with gases and some 'light' liquids such as L.P.G., the pressure drop required to give turbulent flow of liquids is often too high to be attainable. The criteria for laminar and turbulent flow, and their use in selecting line sizes, are discussed in the next section.

8.4.1 Conditions for turbulent and laminar flow

To ensure good mixing and hence a true sample, the flow in the sample line should be turbulent. This is particularly important for gases, and is easily arranged for gas-sample lines, but for liquids it is not usually possible to achieve the necessary pressure drop in the sample line, and laminar flow is generally used instead. The overall analysis time is an important parameter in analyser systems, especially with feedback for computer control of the process, and the time-lag of the sample transport system must not be too long (delay in control) or too short (too high a fast-loop flow is uneconomic and may lead to a process upset). A time-lag of about 30–60 sec is reasonable. The lag can be adjusted by varying the length and size of the line.

The key to the problem is the Reynolds number (R_e), a dimensionless quantity, of the form

$$R_e = dV \rho/\mu$$

where d is the internal diameter of the tubing or pipe, V the linear velocity of flow, ρ the density, and μ the dynamic viscosity, the units used being dimensionally compatible, e.g. d in mm, V in m/sec, ρ in kg/m^3 and μ in centipoise (cP $= \text{g.m}^{-1}.\text{sec}^{-2}$).

If the Reynolds number is greater than about 3000 (some authorities suggest 4000), the flow is turbulent, and if it is less than about 2000 the flow is laminar. There is a transition region between these limits. Thus for a given flow velocity and viscosity, the limiting line sizes for laminar or turbulent flow can be calculated, or the flow velocity for a given viscosity and line size.

The flow velocity is related to the pressure drop in the line by the Darcy equation:

$$h_L = \frac{fLV^2}{2gd}$$

where h_L is the friction head loss, f the friction factor, L the length of pipe, d the diameter of the channel, V the linear flow velocity and g the acceleration due to gravity. Since the pressure drop $\Delta P = \rho h_L$, the pressure drop per unit length of line is

$$\frac{\Delta P}{L} = \frac{fV^2\rho}{2gd}$$

If L is in m, V in m/sec, d in mm, ρ in kg/m^3 and g is 9.81 m/sec^2,

$$\frac{\Delta P}{L} = 51 fV^2 \rho/d \quad \text{mmH}_2\text{O/m}$$

$$= 5 \times 10^{-3} fV^2 \rho/d \quad \text{bar/m}$$

For laminar flow the friction factor is related to the Reynolds number by the expression

$$f = 64/R_e$$

Under turbulent conditions, however, the friction factor is a function of both the Reynolds number and the pipe roughness (defined as the ratio of the surface roughness ϵ to pipe diameter d, both in the same dimensional units). Under *highly* turbulent conditions, f is effectively a function only of ϵ/d. This is shown in the Moody diagram [1] relating f, ϵ/d and R_e. For a fixed value of ϵ, the values of f and R_e can be related to d, as shown in Figs. 8.16 and 8.17. The mean value of the surface roughness (ϵ) of the sample line bore should be about

0.2 μm. It should be noted that care must be taken in using f values, because multiples of f are sometimes used in the literature, but still given the same symbol.

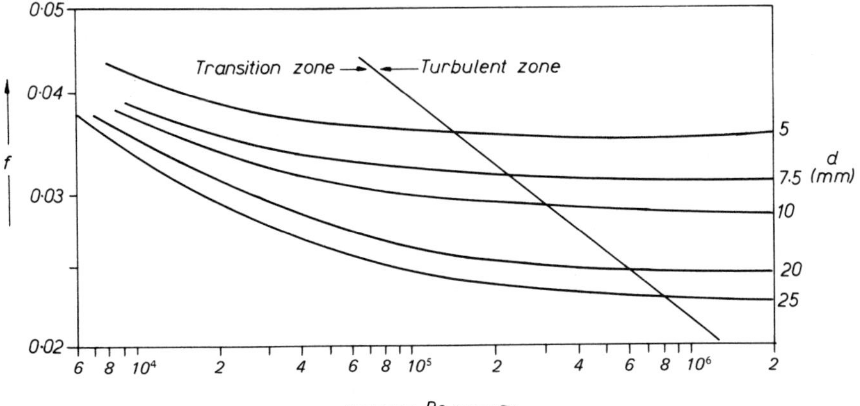

Fig. 8.16 — Variation of Reynolds number with Moody friction factor for different pipe sizes.

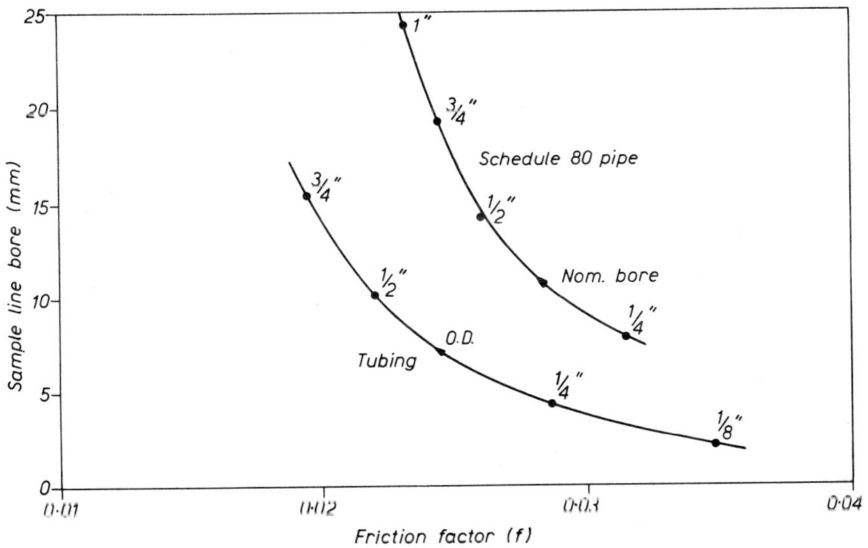

Fig. 8.17 — Variation of friction factor with sample-line bore (for fixed surface roughness, e, with a mean value of about 0.2 μm).

Charts are available (for liquids of density 1000 kg/m^3 and kinematic viscosity 1 centistoke) which give the pressure drop per 100 m line length for various line diameters. For fluids of other densities, the pressure drop obtained from the charts is multiplied by $10^{-3} \times$ density. There are also nomographs available for rapid solution of the equations [2].

If the flow is *laminar*, the relation $R_e = 64/f$ and $V = R_e \, \mu/d\rho$ can be substituted in the pressure drop equation to give

$$\frac{\Delta P}{L} = \frac{fV\rho}{2gd} \times \frac{64}{f} \times \frac{\mu}{d\rho}$$

which on substitution of the dimensions already given yields

$$\frac{\Delta P}{L} = 0.32 \, V\mu/d^2 \quad \text{bar/m}$$

Hence
$$V = \Delta P d^2/0.32 \, \mu L \quad \text{m/sec}$$

and since $\mu = 10^{-3} \, \nu\rho$, where ν is the kinematic viscosity (cSt), ρ is in kg/m³ and μ is in cP,

$$V = d^2 \Delta P/0.32 \times 10^{-3} \nu\rho L \quad \text{m/sec}$$

From this equation, Fig. 8.18 has been constructed for rapid selection of line sizes, based on a Moody friction factor of 0.03 (selected as a working average) and a given pressure drop. Its use is illustrated by the example below, indicated by the arrows on the figure.

$$\text{Sample phase} = \text{liquid}; \; \rho = 1000 \text{ kg/m}^3; \; \nu = 10 \text{ cSt};$$

$$\Delta P/L = 0.033 \text{ bar/m} \, (\sim 330 \text{ mmH}_2\text{O/m}); \; V = 0.6 \text{ m/sec}.$$

Starting from the pressure drop axis on the right, at 330 mmH$_2$O/m (~ 0.033 bar/m), traverse left to the diagonal for a density of 1000 kg/m³, and draw a vertical line through the intersection. Then drop a vertical from 0.6 m/sec on the velocity axis at the top, to cut the diagonal for a viscosity of 10 cSt, and traverse to the right. The intersection of this horizontal with the vertical from the density line gives the sample line size, in this case $\frac{1}{4}$-in. schedule 80 pipe.

Alternatively, we can calculate from the equations given.

$$d^2 = 0.32 \times 10^{-3} \, \rho\nu \, LV/\Delta P$$

$$= 0.32 \times 10^{-3} \times 10^3 \times 10 \times 1 \times 0.6/0.033$$

$$= 58.2 \text{ mm}^2$$

$$d = 7.6 \text{ mm}$$

From Table 8.4 this corresponds to $\frac{1}{4}$-in. schedule 80 pipe. Note that the pressure drop must be expressed in bar/m in the equation, to get the dimensions correct. Since the Reynolds number is given by $Vd\rho/\mu$ and μ (cP) $= \nu$(cSt) $\times \rho$(kg/m³)/10^3, the Reynolds number can also be expressed as $R_e = 10^3 \, dV/\nu$, and in this case is

$$R_e = 7.6 \times 0.6 \times 10^3/10 = 456$$

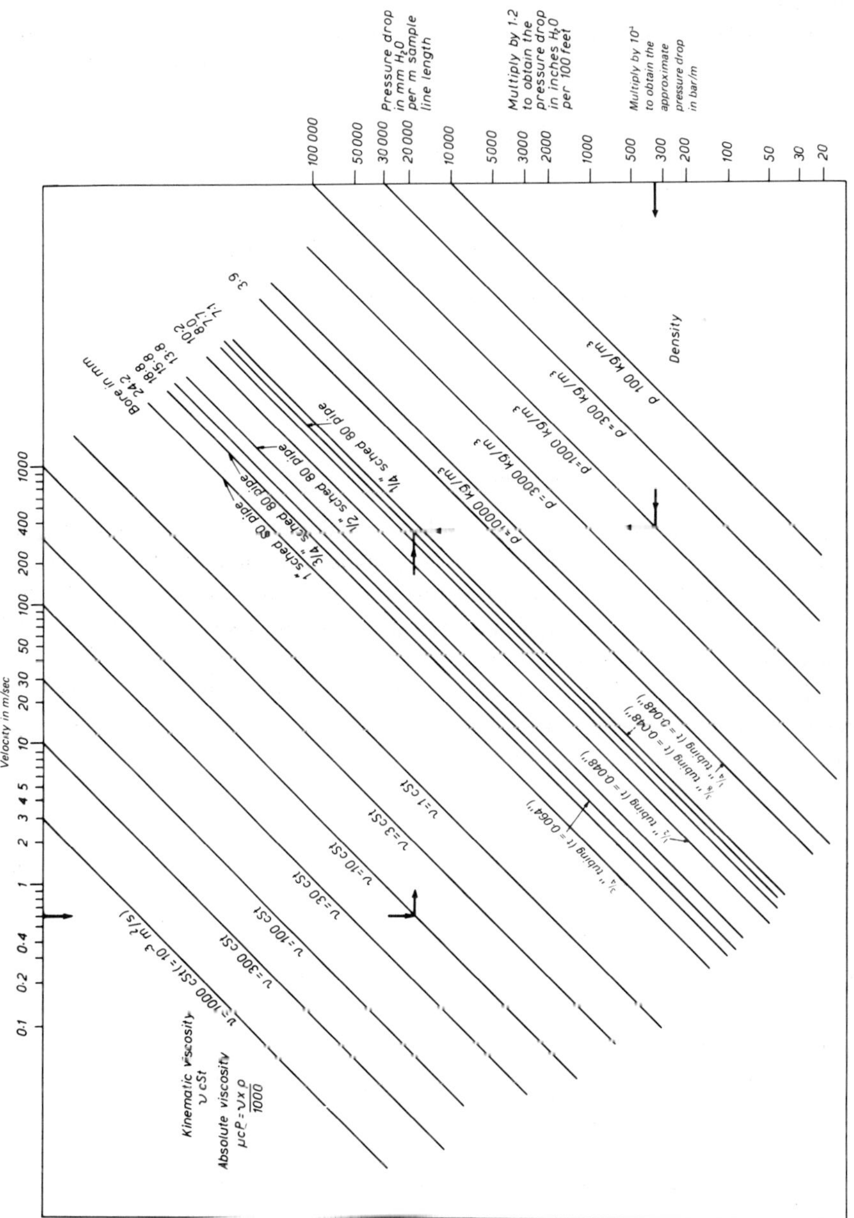

Fig. 8.18 — Chart for selecting sample line sizes.

confirming that the flow is laminar and that the relation $f = 64/R_e$ can be validly used.

For *turbulent* flow, on the other hand, when $R_e > 10^4$, the friction factor may be obtained from Fig. 8.17 and inserted in the pressure drop equation

$$\frac{\Delta P}{L} = 5 \times 10^3 fV^2\rho/d \quad \text{bar/m}$$

to obtain the pressure drop, as in the next example.

Sample = gas: $\rho = 1 \text{ kg/m}^3$; $V = 20 \text{ m/sec}$; $\nu = 2 \text{ cSt}$;

$d = 4 \text{ mm}$ ($\frac{1}{4}$-in. tubing); $f = 0.028$

$$\frac{\Delta P}{L} = 5 \times 10^{-3} \times 0.028 \times 400 \times 1/4$$

$$= 0.0140 \text{ bar/m}$$

$$\sim 140 \text{ mmH}_2\text{O/m}$$

This is fairly close to the drop that would be obtained in practice. The Reynolds number will be $20 \times 4 \times 10^3/2 = 4 \times 10^4$, which is in the transitional range for $f = 0.028$ (Fig. 8.16) and this size of line, but the flow is sufficiently turbulent for the equation to be valid. For a friction factor of 0.04, the pressure drop would be about 0.05 bar/m.

Houser [3] has given a simplified method for sample line sizing, based on curves similar to those shown in Fig. 8.19. One set of curves is supplied for each of a number of typical sample line sizes.

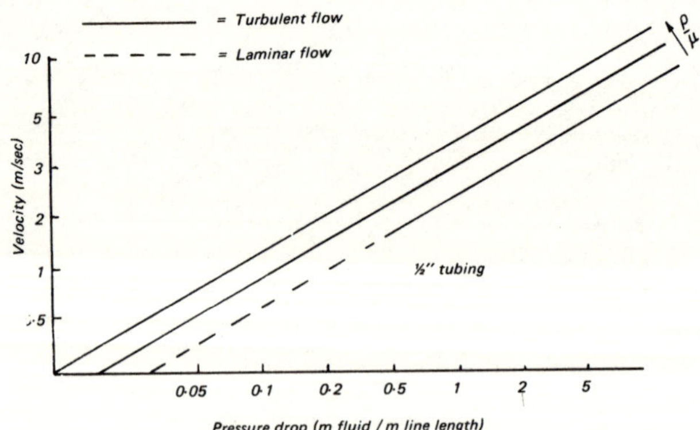

Fig. 8.19 – Sample handling curves.

To determine whether the size of a sample line is sufficient, the Reynolds number is first calculated to see whether the flow is laminar or turbulent, and which equations will be applicable. The flow velocity is then calculated from the appropriate pressure drop equation, and checked to see whether it is attainable, and sufficient for the required transport lag t:

$$t = V/L \text{ sec}$$

The line size may be increased if necessary.

This applies to long lines where $L/d > 10$ and there is sufficient mixing. If $L/d < 10$, multiply by a mixing coefficient which could be as high as 3. This applies only to relatively incompressible fluids, and to compressible fluids where $\Delta P/\bar{P} < 20\%$, and the average pressure $\bar{P} \cong 0.9 \times$ inlet pressure. If $\Delta P/\bar{P} > 20\%$, the flow may be appreciably compressible. It may be assumed to be isothermal if $L > 30$ m. For more details see Lobo *et al.* [4].

For compressible flow, the flow F_s under standard conditions can be calculated from the actual flow F at absolute pressure P (bar abs) and temperature T (K):

$$F_s = \frac{FPT_s}{P_s T}$$

The contribution of bends, unions, valves, strainers etc. to the line pressure drop is usually small enough to be omitted in estimation of the line size.

Pressure drops in most sample-conditioner components are determined and included in the loop calculations. They can be estimated with the help of Table 8.5, but the values given are very approximate and will vary considerably according to the components used and the nature of the sample.

Pipe is thick-walled and is supplied in short straight lengths. Pipe size is defined by its nominal bore and schedule number (a wall-thickness description). Pipe ends are joined to fittings by welding or screwing. The minimum size commonly used is about $\frac{1}{4}$ in., the schedule number then being typically 80.

Tubing is relatively thin-walled and is often supplied coiled in long lengths. Tubing size is defined by its outside diameter (o.d) and wall thickness. Tubing ends are joined by compression fittings. The maximum size commonly used is about $\frac{1}{2}$ in. It is often preferable to use 37° flare tube fittings for connection to copper tubing.

Metallic tubing should be seamless, soft, bright annealed and cold drawn. Typical materials used are listed in Table 8.6. Drawing is sometimes done with a special grease rather than soap, since it is easier to remove from the open pores.

Tubing sizes commonly used are shown in the shaded areas of Table 8.7.

Table 8.5

A guide to typical pressure drops across components at the normal flow rates

Component	Condition of sample	Sample fluid	Pressure drop, bar	Remarks
Sample line		Liquids	2–3 per 100 m	Laminar flow
Sample line		Gases	1–1.15 per 100 m	Turbulent flow
Strainer	clean	Liquids	0.02	
	dirty		0.07	
Cyclone separator		Liquids	1.5–2	
Filter	clean	Liquids	0.03	
	dirty		0.15	
Filter	clean	Gases	0.05–0.1	
	dirty		0.7	
Oil filter/coalescer	clean	HC liquids	0.07	
	dirty		0.7	
Gas filter/demister	clean	Gases	0.05–0.1	
	dirty		0.3	
Restriction orifice for constricting flow		Gases	50% of abs. inlet pressure	Critical flow
Needle-valve for flow regulation		Liquids	1.5	
Needle-valve for flow regulation		Gases	50% of abs. inlet pressure	Critical flow
Flow controller consisting of pressure-control valve and needle-valve		Liquids	0.7–1.5	
Flow controller consisting of pressure-control valve and needle-valve		Gases	0.7–1.5	
Variable-area flow-meter		Most	0.02	

Table 8.6

Materials used for metallic tubing and pipe

Material	Tubing or pipe	Specification	Further description
Aluminium	Tubing	ASTM B241	Alloy 6061–T6
Copper	Tubing	ASTM B75	
Carbon steel	Pipe	API 5L	
Carbon steel	Pipe	ASTM A161	For high temp. service
Alloy steel	Pipe	ASTM A335	Ferritic. For high temp. service
Stainless steel	Tubing	ASTM A268	Ferritic. Grades TP405, 409 etc.
Stainless steel	Tubing	ASTM A269	Austenitic. Grades TP304, 304L, 316, etc.†
Stainless steel	Tubing	ASTM A270	Grade TP304, with special surface finish
Stainles steel	Pipe	ASTM A312	Austenitic. Grades TP304, 304L, 316, etc.†

† L = low carbon.

Table 8.7
Commonly used tubing sizes

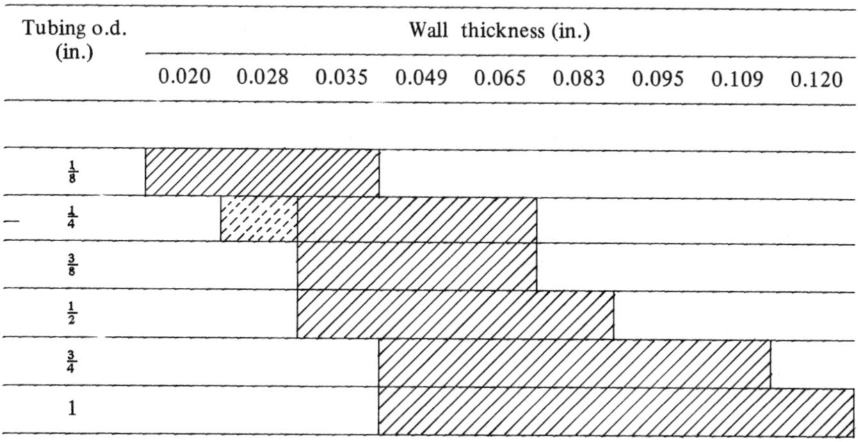

Tubing o.d. (in.)	Wall thickness (in.)								
	0.020	0.028	0.035	0.049	0.065	0.083	0.095	0.109	0.120
$\frac{1}{8}$									
$\frac{1}{4}$									
$\frac{3}{8}$									
$\frac{1}{2}$									
$\frac{3}{4}$									
1									

Upstream of the main filter, the material of the fast sample loop could be the same as that of the process line. The material of the remainder of the sample line should be stainless steel unless it is not suitable for the sample fluid.

In order to minimize the number of fittings in sample lines, tubing and pipe should be used in long lengths. Small-size tubing should typically be supplied coiled in lengths of 15–50 m. Pipe should be supplied in straight lengths of between 5 and 10 m.

8.4.2 Care of the sample line

Surface adsorption is minimized if the surface area is minimized by using the smoothest economically available finish; bright annealed tubing has a good surface finish.

The inside of the sample line should be smooth and clean. Since it is very difficult to polish small-bore tubing or pipe, it is necessary to select carefully drawn material. The use of fine carborundum powder and compressed air for polishing has been proposed but has not been very successful for stainless steel which contains large crystals. Burnishing would be more effective if it were possible.

Sampling errors due to adsorption and reaction at the interior surface of sample lines are minimized by making the contact area as small as possible.

Obviously a long sample line can behave as a chromatographic column and produce undesirable unequal transport lags if precautions are not taken.

If the line is dirty it should be cleaned. A typical cleaning sequence is as follows.

(i) Flush with acid at a low flow-rate. Hydrochloric or phosphoric acid is often used. For more vigorous cleaning a mixture containing 20% nitric acid and 2% hydrofluoric acid may be used.

  (ii) Flush with clean water at normal sample flow-rate to remove the acid and salts.

 (iii) Flush with alcohol at low flow-rate to remove the water.

 (iv) Flush with acetone at low flow-rate to remove the alcohol.

  (v) Pass warm dry nitrogen to remove the acetone.

The line should then be sealed. Steps (iii) and (iv) can be replaced by a single flush with a solvent such as I.C.I. "Arklone".

Sometimes an acid wash is not required but a solvent wash is necessary. Various solvents are used, such as aromatics (benzene and toluene) or chlorinated hydrocarbons (methylene chloride and trichlorethylene). The solvents must be fresh.

Freon solvents (Du Pont) may be used for cleaning lines after the water wash. The water may be displaced by Freon TP 35 which contains 35% of iso propyl alcohol; this should then be followed by Freon TF which has a boiling point of $47.6°C$ and thus may be removed with nitrogen at a temperature above this.

If it is not necessary to use an acid wash. Freon TMC may be used. This contains methylene chloride and has a boiling point of $36.5°C$, thus being removable by a hot nitrogen flush. PTFE tubing may also be cleaned with these Freon solvents. These more expensive solvents may be recovered in a separator where the water, dirt and other impurities are removed. The use of a more effective solvent decreases the labour and time spent in cleaning.

Unless proper precautions are taken, this cleaning effort may be wasted by passage of a dirty sample, especially if the sample is obtained from a unit which is in the early stages of commissioning.

It is often sufficient to flush the line with a relatively clean process sample for several days. During the first stage, the analyser should be isolated from the sample line and the analyser may be commissioned and the calibration checked with the calibration fluid. During the second stage the analyser may be fed with the process fluid. When the reading has stabilized and the reproducibility is satisfactory, the sample loop should be ready for acceptance.

Frequent washing followed by stabilization with the normal sample stream reduces the rate of build-up of more difficultly removable deposits and increases the long-term availability of analysers, especially those fed with samples that contain high boiling-point components or that polymerize.

When the analyser is out of commission it is advisable to isolate the sample line and to purge it with dry, clean, oxygen-free nitrogen at a low flow-rate.

Condensation can occur if the vapour sample temperature is too low; the result can be line wall corrosion if the wet vapour is acidic, and fractionation if important heavy components condense. It should be prevented by the use of steam or electrical tracing (i.e. heating), but the temperature must not be made too high, or pyrolysis, polymerization or coking may occur. The maximum surface temperature must be limited according to the temperature class of the gas which could leak into the surrounding atmosphere. Class T3 limits this temperature to 200°C. Steam heating is generally used, though it is not so easy to control as electrical tracing, but the latter must be installed properly to ensure that it is safe. On off-shore platforms, electrical heat tracing is often preferred. The temperature of the gas sample should always be above the dew-point temperature. Overheating increases the rate of polymerization, and for this reason olefin sample lines are often run in pairs, with one as a spare. Hydrates (clathrates) can be formed when hydrocarbon vapours containing free water are at high pressure and low temperature, and can rapidly clog narrow nozzles and sample lines, and again heat tracing should be used.

When 'Dekatrace' (Eaton) heat-traced sample lines are used, the upper and lower temperature limits for a particular line length are calculated and the desired range is achieved by use of calculated thicknesses of thermal insulation, and tracing. For electrical heat tracing, a self-limiting product such as 'Chemelex' (Raychem) may be preferred, provided that the heat supplied is sufficient (typically 33 W/m).

Heat-traced lines should always be insulated thermally. Seldom is it necessary to apply a thermally conducting cement.

If the sample corrodes type 316 stainless steel, a super-stainless steel such as Carpenter No. 20 Cb3 may be more resistant. Alternatively, titanium or a suitable plastic such as nylon, PTFE or polypropylene may be used. Viton tubing may be used inside equipment. Plastic-lined metal tubing is also available. However, plastics have temperature and pressure limitations.

If necessary, the inner surface of metal sample lines may be silanized to reduce adsorption of water and unsaturated hydrocarbons. Water is very polar and tends to be adsorbed on the surface of metals, so the surface area of the line in contact with the gas should be minimized. If a very low concentration of water is to be measured, it is best to use plastic tubing with walls that are not porous to water. The best materials appear to be high-density polyethylene and oriented polypropylene.

In sampling wet acid gases it is important to minimize the reaction of the gas with the sample line. When the acid gas is wet H_2S, nylon should be used since it is not very porous to H_2S and is not attacked by it. Stainless steel is corroded by wet H_2S, but aluminium and titanium are more resistant. Aluminium should not be in contact with a saline atmosphere. Also, it is not used in oil and gas plants. When it is used, an alloy with a low magnesium content is preferred. When a low concentration of H_2S in the gas is to be measured, free water should be removed at the sample take-off point.

The following check list is useful for optimum selection of a sample line

Sample Line Calculation and Selection Sheet

Tag number

Take-off line material

Polishing and cleaning

Return line material

Cleaning

Fluid

Phase

Density

Viscosity

Take-off pressure

Return pressure

Max. allowable take-off line pressure-drop

Max. allowable return line pressure-drop

Take-off line length (L_1)

Return line length (L_2)

Take-off line pressure-drop

Maximum allowable transport lag time (T_m)

Minimum velocity (L_1/T_m)

Pipe or tubing

Line size (nominal bore of pipe or o.d. and wall thickness of tubing)

Actual bore

Volume per unit length

Volume flow-rate (linear velocity × volume per unit length)

Friction factor (f)

Type of flow required (laminar or turbulent)

Reynolds number

Selected velocity

Resultant lag

Selected flow-rate

Return line pressure-drop

8.5 SAMPLE CONDITIONERS

By-pass sample loops (Fig. 8.20) must be used whenever possible so that the sample can be returned to a convenient point in the process unit. This is particularly important for liquid streams. If it is not possible with vapour streams, the sample may be vented to atmosphere [Fig. 8.20 (*b*)].

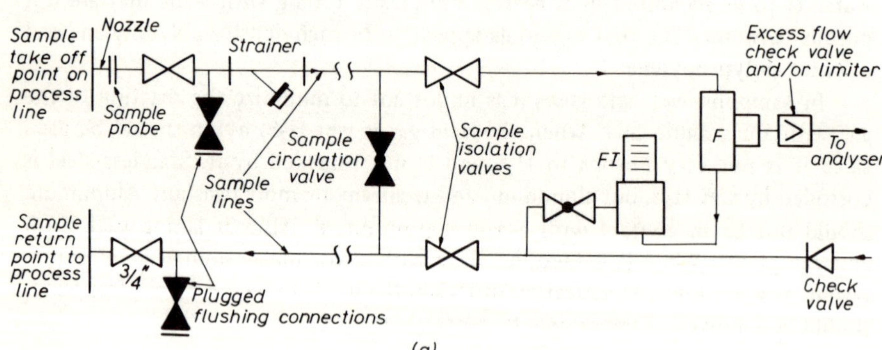

Fig. 8.20 (*a*) – By-pass sample loop.

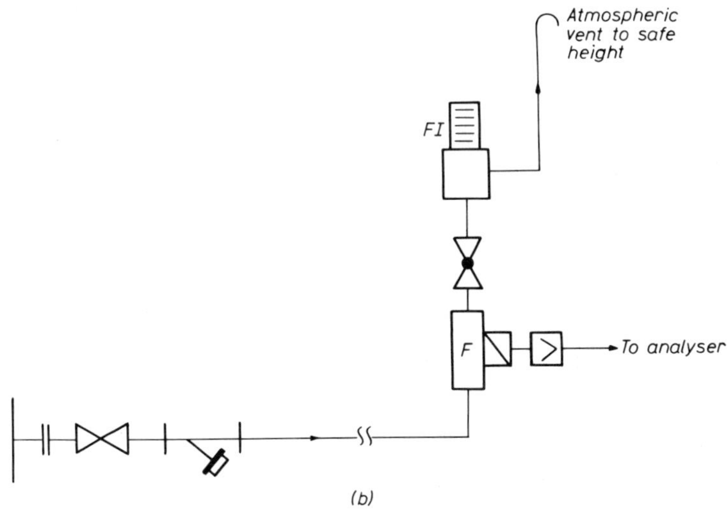

Fig. 8.20 (*b*) – Sample system with the by-pass vented to atmosphere.

Isolating valves at process lines should normally be solid wedge-gate valves. Other isolating valves within the analyser systems should be ball valves. Valves and most other components should have female screwed connections.

The by-pass fast sample-loop variable-area flowmeter ought to be a stainless-steel armoured meter; a glass meter should not be used for this location.

If the analyser is on control or interlock duty, a flow-switch should be fitted to the flowmeter. This should operate a sample-failure annumciator in the analyser house and in the control room [Fig. 8.21 (*a*)]. A pressure-switch may take the place of the flow-switch.

The amount of sample entering the analyser house must be limited. Thus the by-pass filter should be located outside the analyser house and a flow-limiting restriction should be fitted; this could be a needle-valve or a pressure-reducing valve. For a vapour sample, the outlet pressure should not exceed half the inlet pressure, so that critical flow exists.

Use may be made of the fast by-pass loop flow rate to provide motive power for an ejector to return low-pressure samples to process, instead of to drain, which always presents hazard problems, particularly if the sample is hot and has a high vapour pressure [Fig. 8.21 (*b*)].

Suitable ejectors are available from Körting (Germany) and Penberthy and are often supplied with Totco boiling-point and other analysers. Care must be taken that the ejector is well polished and of proper size. Alternatively, the sample may be returned to a pressurized slops system, or to process through a sample recovery unit (Fig. 8.22).

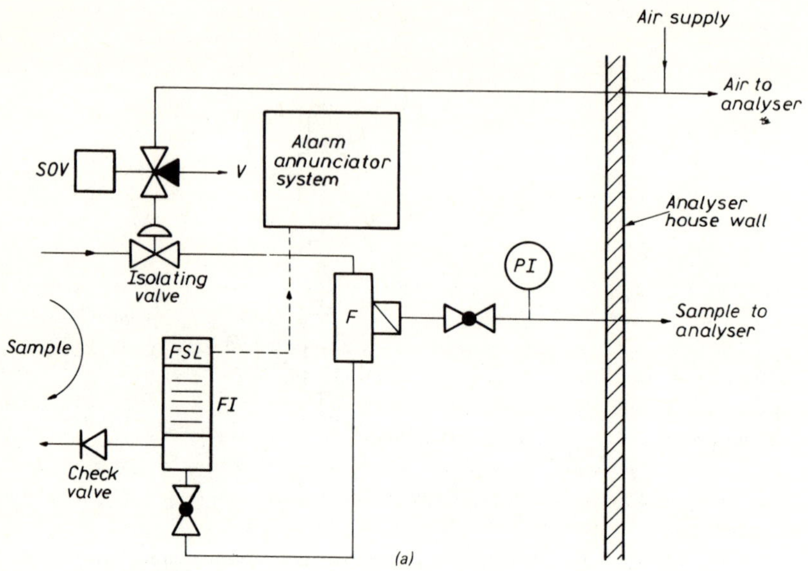

Fig. 8.21 (*a*) -- Sample-flow failure annunciation and safety isolation.

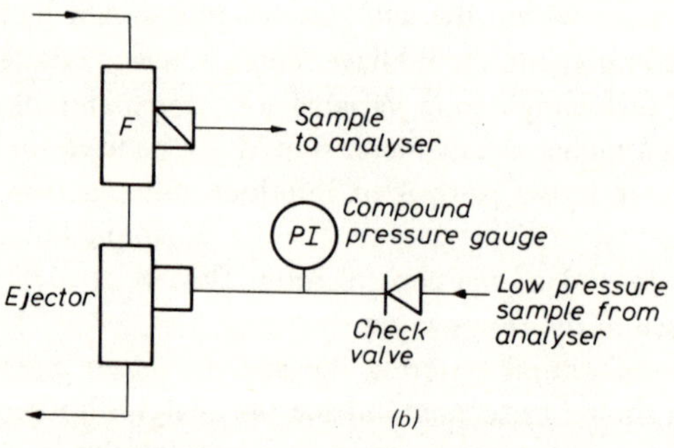

Fig. 8.21 (*b*) – Use of an ejector.

8.5.1 Filters

Most samples contain rust particles picked up from process lines and equipment. Some samples contain wet H_2S which reacts with iron to form iron sulphide particles. Catalyst fines appear in some samples, especially from catalyst regeneration stack gas. It is thus advisable that all analyser systems should be fitted with a strainer, close to the sample take-off point, to protect the downstream components from damage and blockage due to the larger particles [Fig. 8.23 (*a*)]. The mesh size selected should be as coarse as possible, to minimize the pressure drop.

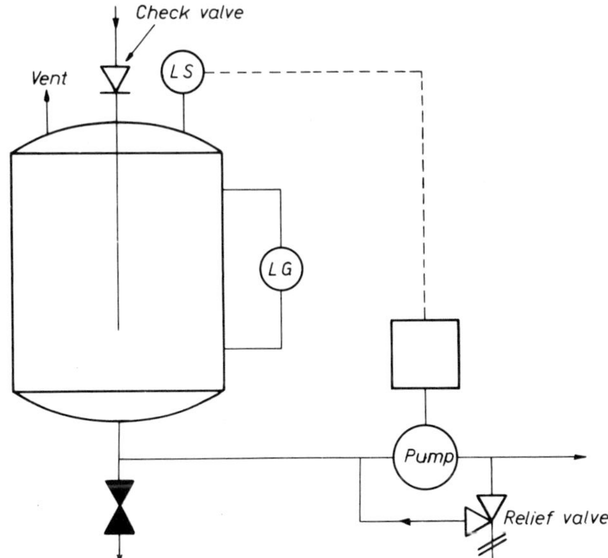

Fig. 8.22 – Sample recovery unit.

Particles tend to separate out from the main stream and concentrate near to piping walls, so the concentration of particles in the sample stream will be much reduced if the sample is taken from the centre of the pipe rather than from near to the wall.

It is standard practice to use at least two filters. A by-pass filter in the fast loop with a 25–100-μm rating should be followed by one or more 2–5-μm rating series guard filters. The filter may be a diaphragm filter, a depth-type filter, or an edge-type filter.

Depth-type filters may be sintered stainless steel with pore sizes down to about 10 μm, porous ceramic with a pore size down to about 0.5 μm, or inert fibres. The finer filters tend to clog easily and thus to produce a large pressure drop. For this reason, the surface area should be increased. A supported thin flat fibrous membrane filter is sometimes preferred.

A Doulton Aerox G28 ceramic filter contains pores with a mean diameter of 60 μm and retains 20-μm particles. If this element is found to be too coarse, the by-pass filter pore size could be increased to 100 μm or even larger and a finer pore-size element added after the by-pass filter.

The pressure drop across a filter should be limited so that an adequate flow-rate is obtained and the bursting limit is not approached when the element is dirty. A differential pressure indicator is sometimes fitted across a filter, or sometimes a pressure gauge and selector valves are used [Fig. 8.23 (*b*)].

An alternative system uses a filter in the fast and slow loops. As a result, the fast-loop flow indicator and flow-regulating valve are kept clean. The valve,

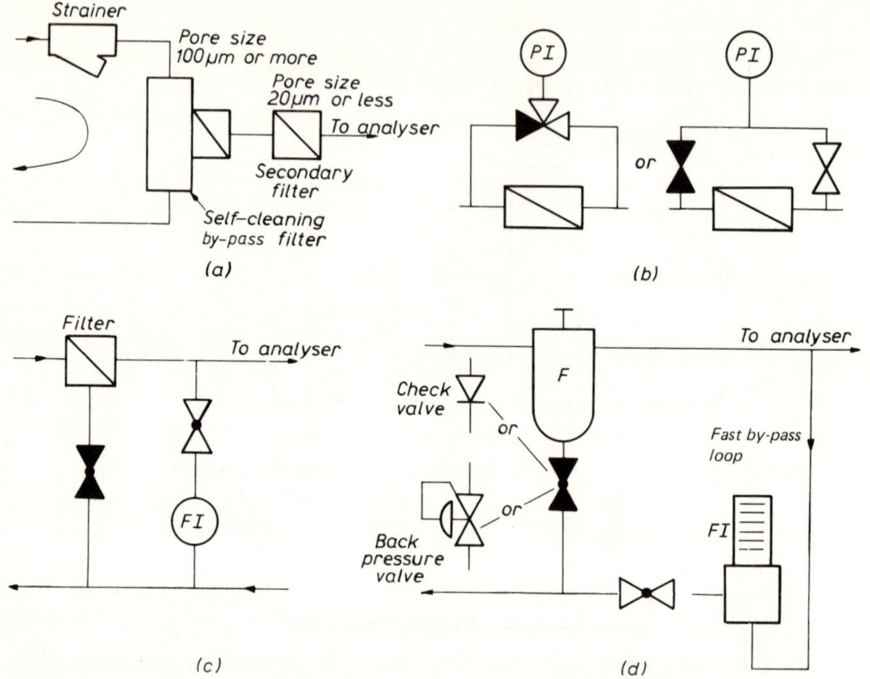

Fig. 8.23 — (a) Filter arrangement. (b) Measurement of differential pressure across strainers. (c) Filter cleaning. (d) Use of a check valve or back-pressure valve.

which is normally closed, is used when the filter is cleaned [Fig. 8.23 (c)].

The flow through the by-pass filter should be at high velocity and downwards, so that the vertical filter-inlet surface is swept clean and the dirt falls out of the filter. It is important to ensure that fractionation does not occur. In chromatographs such splitting has to be very carefully controlled.

For gas samples, self-cleaning centrifugal membrane filters are often used. Examples are the Collins Swirlklean and the Beckman Kleanstream filters. The membrane element has a typical pore size of 0.5 μm and is available in a range of pore sizes from 0.3 to 10 μm. The Collins Swirlklean filter is also available with a coalescing element to reduce the moisture content in the filtered sample.

When waxy oils are to be sampled, the waxes should remain in solution as they can soon clog a filter, so the temperature of the sample in the filter should be as far above the cloud point as practicable.

When fine particles or dusts are present, the number of particles of any particular mesh range usually increases considerably as the size decreases. The smallest ones usually remain in suspension and do not coagulate, owing to the repellent action of their electrostatic charges. The smaller particles are difficult to remove, but, if they do not agglomerate, there should be no problem of blockage of orifices.

These residual small particles may, however, cause problems by accumulating on the analyser element (e.g. that of the aluminium oxide hygrometer) where they block sites and hence reduce the analyser sensitivity.

Edge-type filters, such as the Cuno Turnoklean and the Plenty Clinsol, are very suitable for use as primary filters with liquid streams, because cleaning is easily done without back-flushing and without element removal, by rotating the rotor shaft, typically a few turns every day or so, according to the amount of dirt present. While the filter is being cleaned the bowl should be emptied. The rate of emptying should be limited in order to ensure that the analyser sample supply is not severely restricted. To limit analyser sample-starvation when a continuous drain is required, the by-pass return valve may be replaced by a check valve or a back-pressure valve [Fig. 8.23 (*d*)]. Edge-type filters are coarse, so they should usually be followed by a fine filter.

The normal Cuno element spacing is from 0.020 in. (0.51 mm) to 0.0035 in. (0.09 mm), equivalent to 35–170 mesh, removing particles of 500–100 μm breadth. The Super Turnoklean filter has a second stage which can provide filtration down to 40 μm with element spacing of 0.0015 in. (0.038 mm).

It is possible to buy a timed pneumatic drive to rotate the cleaning shaft of the Cuno filter, to give 1 complete turn in 10 sec, at 3–4 hr. intervals.

Cyclone separators remove solid impurities from liquid sample streams by means of centrifugal action followed by momentum separation. The clean liquid tends to concentrate in the centre of the vortex and then changes direction rapidly before it leaves the separator. Solid particles more dense than the liquid have more momentum and thus do not change direction so easily as the liquid, so they flow out of the bottom of the separator.

When the sample is very dirty, cyclone separators [Fig. 8.24 (*a*)] should be used upstream of the primary filter to reduce filter cleaning. Cyclone separators are available from the mechanical seals division of Borg Warner and from Crane Packing. MSA make a vortex type separator for gas samples.

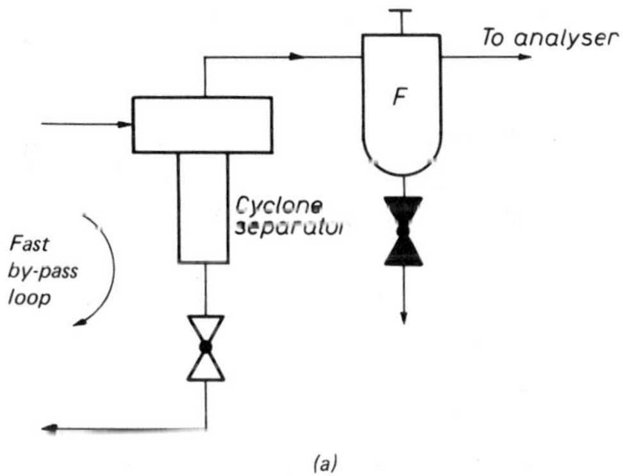

Fig. 8.24 (*a*) – Use of a cyclone separator.

Liquid filters always remove small particles more effectively than dry filters, but they cannot always be used, for the liquid could easily absorb an important component of the sample or could contaminate the stream with a significant impurity. "Scrubbing" through a liquid filter can be advantageous for the removal of corrosive components which are not required to be measured.

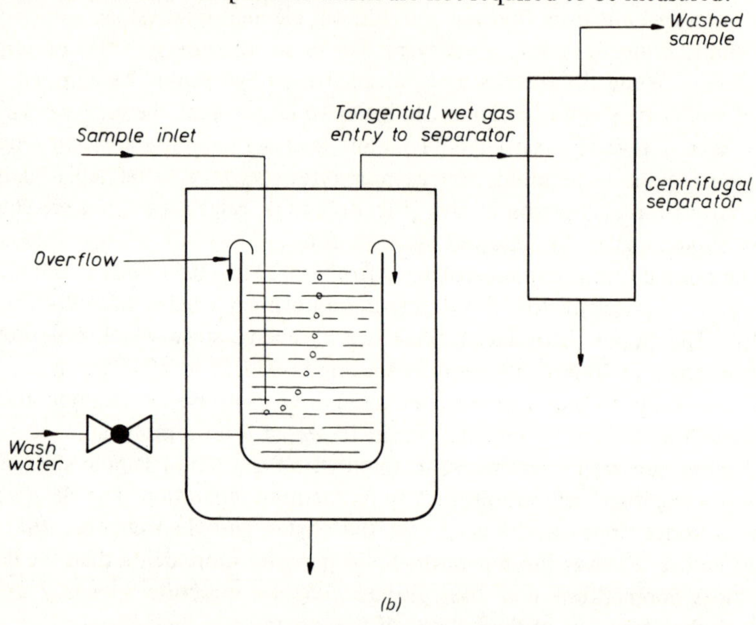

Fig. 8.24 (b) – A typical wet gas cleaning system.

Figure 8.24 (b) illustrates a typical wet-gas cleaning system. The dirty gas is first bubbled through a fresh stream of wash water. The water suspended in the cleaned gas in then removed in a centrifugal separator [Fig. 8.24 (c)]. A more complete scrubbing system is shown in Fig. 8.24 (d). Absorption-type driers have not been included since they often introduce an unacceptably high residence time.

A dirty sample which contains a condensible vapour or which is wet should preferably be filtered at a temperature higher than the dew-point temperature. After filtration, the sample can be cooled and the condensate removed, Fig. 8.24 (e).

Magnetic separators are sometimes used to improve the removal of magnetic particles from sample streams.

8.5.2 Coalescers

Coalescers are used to remove free water from liquid hydrocarbons and to perform secondary filtration to, say, 2 μm. They are only suitable for light hydrocarbons not heavier than gas oil, beyond which coalescing is not very efficient. According to ASTM D86, the sample is first cooled to 15°C. If it

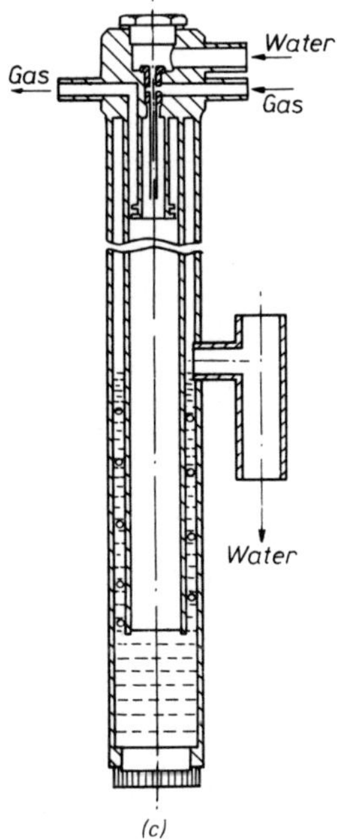

Fig. 8.24 (*c*) — A centrifugal separator.

contains a visible quantity of free water, the sample is discarded if the expected boiling point is less than 66°C. Thus coalescers are fitted to boiling-point analysers not measuring end-points. They are also fitted to low-temperature analysers such as those for cloud-point, pour-point, and freezing-point measurement and to analysers measuring the absorption of light beams, such as colour analysers and infrared analysers. Coalescers are made by Hallikainen, Selas and Fram.

Coalescing causes the fine droplets of the dispersed phase to conglomerate and to form a clearly defined phase. The greater the interfacial tension, the more readily will the dispersed droplets coalesce. Typical values of the interfacial tension of various hydrocarbon liquids containing water droplets are shown in Table 8.8.

Table 8.8

Interfacial tensions of various hydrocarbon liquids and water.

Hydrocarbon	Interfacial tension, in dyne/cm $(10^{-3}$ N/m) at about 20°C
Benzene	35
Gasoline	48
Gas oil	62

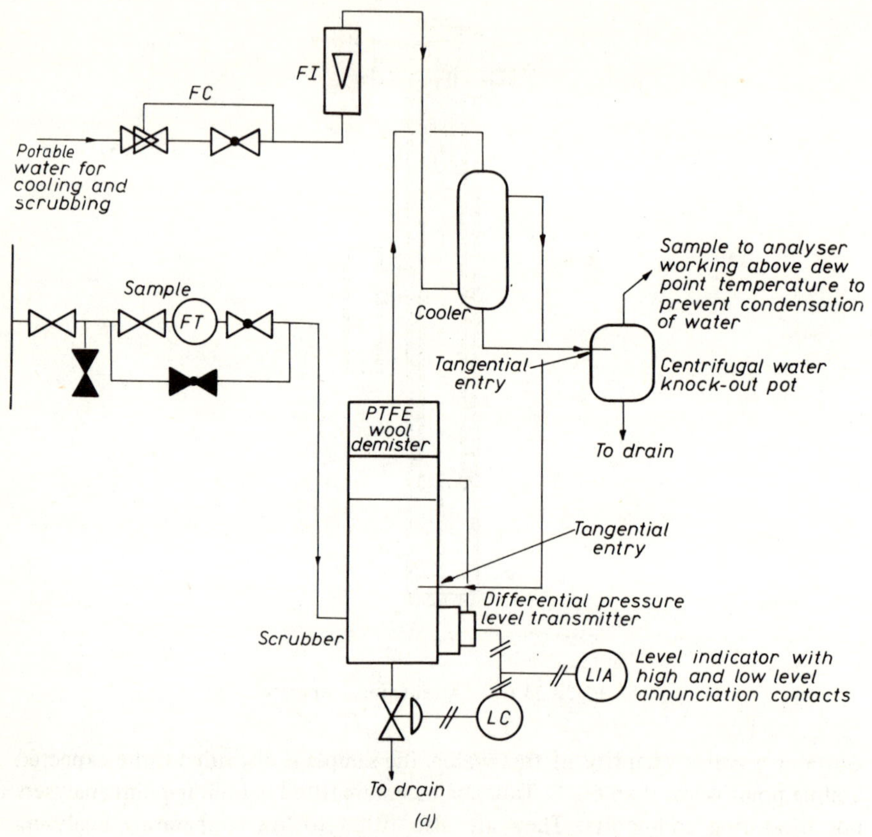

Fig. 8.24 (*d*) – A scrubbing system.

Surface activity, fine dust particles in the emulsion and a viscous continuous phase reduce the efficiency of coalescence.

The permanence of a dispersion results from the surface of the dispersed droplets being electrically charged or covered with a film of surface-active agent, which prevents conglomeration of the droplets on collision.

After coalescence, the two phases are separated by settling according to density difference. For this purpose time is required. The settling section should be designed so that the velocities of the liquids are low enough for adequate settling to occur. Selas Flotronics and Jowa of Belgium make several types of coalescers.

(i) *Surface coalescers* consist of Teflon-coated porous porcelain or stainless-steel cylinders or plates. These are used for liquid–gas separation, i.e. demisting.

(ii) *Depth coalescers* are made from fibre-glass elements. They are used for liquid–liquid separation in which the drops coalesce inside the fibre-glass pack.

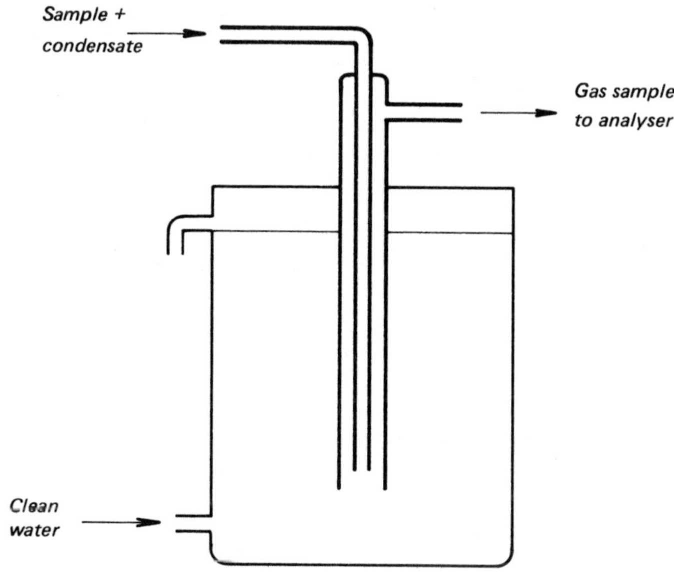

Fig. 8.24 (*e*) – A condensate trap.

Various grades are available – ultra-fine, fine, medium and coarse. The choice of grade is based on the emulsion viscosity and surface tension. A liquid will only flow through fine capillaries if it preferentially wets them; if the pressure drop is too high, a capillary-type membrane can burst and, as a result, leakage will occur.

For the best efficiency, a low velocity of, say, 10 m/hr (to give long contact time) and a controlled differential pressure of the order of 0.2 bar should be used. The thickness of the membrane is selected according to the difficulty of separating the emulsion into its two phases.

Pyrex glass bodies are available for laboratory use up to 10 bar ga. For process applications, stainless-steel bodies should be used and these should be fitted with an interface gauge glass if the coalescer is not too small. The effectiveness of fibre-glass as a coalescer is due to the fact that, like water, it is polar and thus retains water preferentially.

Other packings developed are made of polypropylene and Teflon (Du Pont). Stainless-steel wool is also useful.

The location of the coalescing membrane in its container depends on whether the lighter or the heavier phase is required to be analysed (Fig. 8.25).

A single-stage coalescer can separate typically 98–99% of the free water from a wet hydrocarbon liquid feed. This is sufficient for most applications. A three-stage coalescer (Fig. 8.26) can remove about 99.5% of the free water.

Before coalescing, the liquid sample (emulsion) should be passed through a strainer to remove large particles, a filter to remove dust, which would clog the

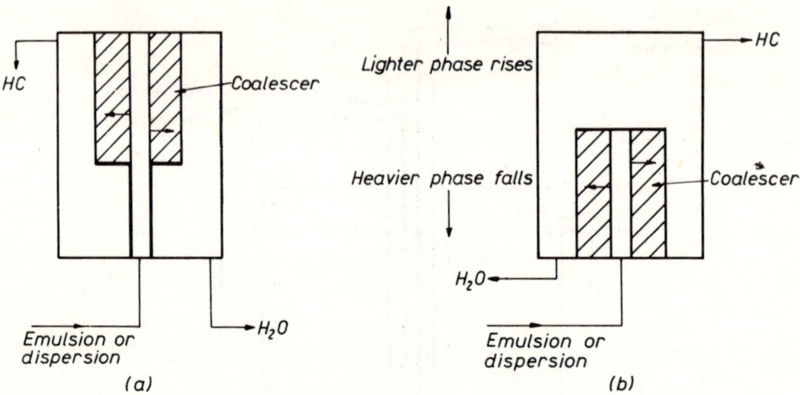

Fig. 8.25 — Two positions of the coalescer in its container (HC = hydrocarbon).

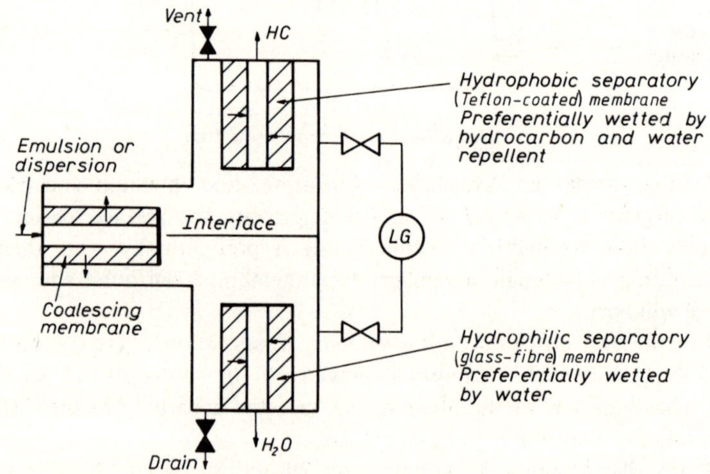

Fig. 8.26 — Three-stage coalescer.

coalescer, and cooled.

If the sample is cooled after coalescing, more free water is released. An activated-carbon absorber may be used to remove traces of oil from a water sample. A coalescer and absorber would thus be useful for the removal of oil from a sour water sample before pH measurement, thus reducing the amount of oil deposited on the glass electrode and making the measurement more reliable (Fig. 8.27).

A coalescing filter is incorporated into many liquid-sample conditioning systems to remove the water in suspension (Fig. 8.28). Coalescers are most efficient when working with cold clean samples. Finely divided solids tend to accumulate at liquid interfaces and, as a result, coalescence is retarded. In the

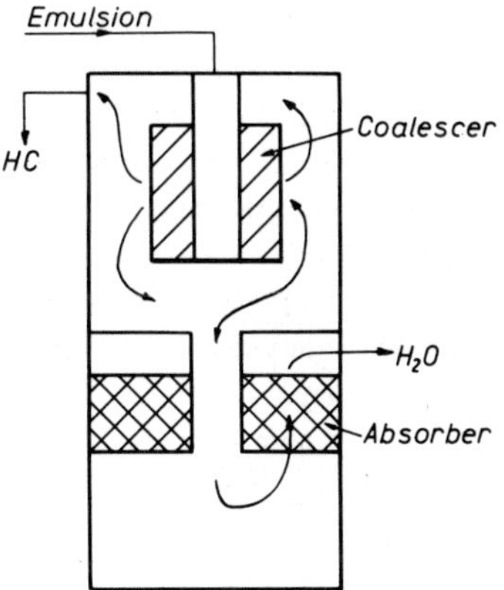

Fig. 8.27 – Combined coalescer and absorber.

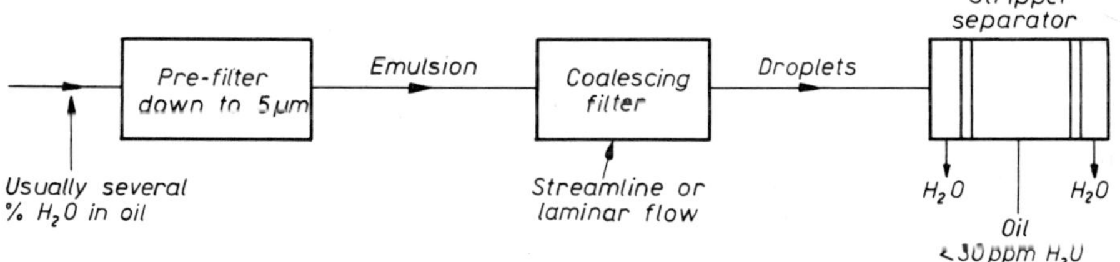

Fig. 8.28 – Coalescing filter system.

Fram coalescing filter the water may be fed to drain through an automatic dump. Coalescing filters will not remove the water in solution. For this reason, the temperature must be kept low [Fig. 8.29 (a)] .

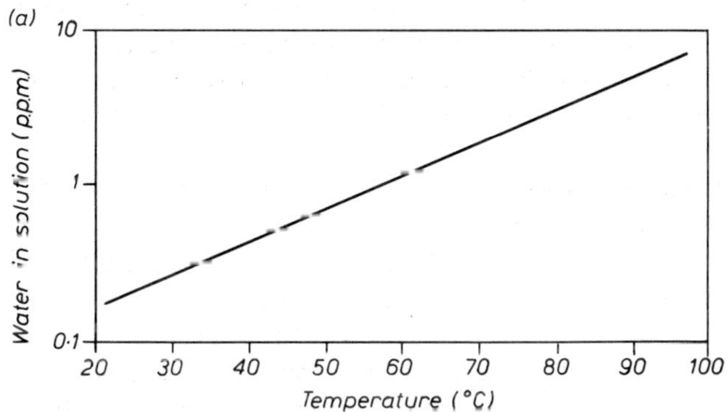

Fig. 8.29 (a) – Variation of water in solution with temperature.

Streamline or laminar flow promotes coalescence. Turbulent flow produces emulsification. For coalescence to be fast, the interfacial tension must be large. Surfactants must not be present; otherwise, coalescing could be impossible. Some filming amine corrosion inhibitors reduce the surface tension of the sample so much that coalescing is impracticable and other methods of water haze-point reduction have to be used.

The water draw-off [Fig. 8.29 (*b*)] usually requires frequent adjustment and it is strongly recommended that analysers be visited regularly, even daily, to ensure that the water draw-off is correct.

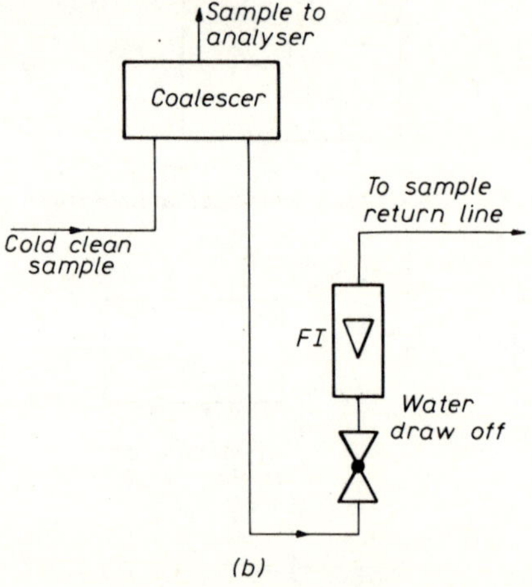

Fig. 8.29 (*b*) – Water draw-off.

To ensure removal of the separated water from the coalescer, the water must drain freely from it. The coalescer should thus be located as high as is convenient (Fig. 8.30) and a minimum amount of restriction should be in the water draw-off line. It is thus best not to include a needle valve in this line. Removal of the variable-area meter would also remove a source of water trapping. The inlet flow-rate should be adjusted to about three times the analyser-sample flow-rate. Coalescers may also be used to remove oil from the water feed to a pH cell (Fig. 8.31).

Salts are more soluble in water than in hydrocarbons, so the haze-point may be reduced by passing the sample through a sodium chloride or other salt bed. This has to be replaced as it becomes spent. A haze-point below about $-5°C$ is usually difficult to obtain by such methods.

Alternatively, molecular sieves may be used, arranged in paired beds so that one may be regenerated by heating while the other is being used.

Air blowing can also be used, provided that sufficient air can then be removed from the liquid sample before it is passed on to the analyser. Blowing should be done at as high a temperature as the analysis permits: cooling after blowing may therefore be necessary.

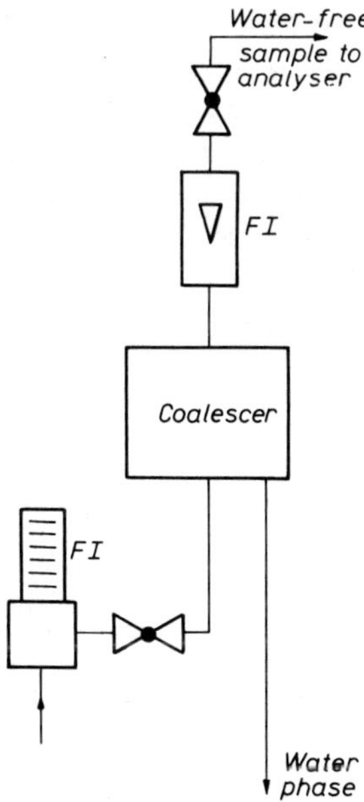

Fig. 8.30 – The correct installation of a coalescer.

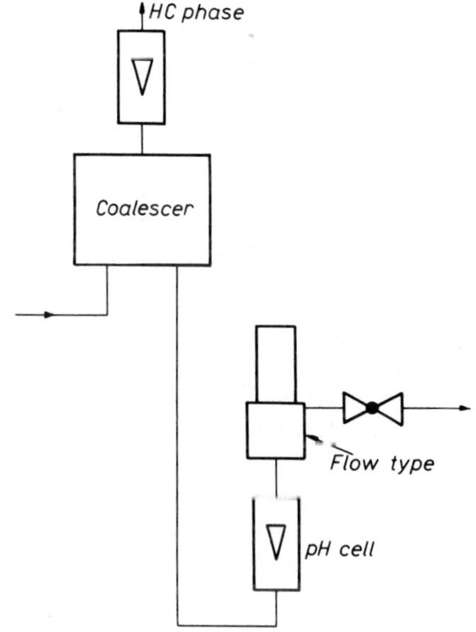

Fig. 8.31 – Coalescer used to remove oil from a water feed to a pH cell.

8.5.3 Other separators

Hone supply a steam-heated air-blowing water stripping tank (Fig. 8.32) for the effective removal of free and dissolved water from gas oil and other similar liquids. They supply it with their cloud-point analyser.

Demisters should be used to remove suspended fine liquid droplets from gas samples. They should be fitted to gas analyser systems where the sample temperature is below the dew point. The surface of the demister should be inert and should be easily wet by the droplets, so that it promotes coalescing of the mist into larger droplets, which drop down into a self-emptying pot [Fig. 8.33 (a)] .

PTFE wool has been used successfully and it is preferable to glass fibre wool which can produce dust.

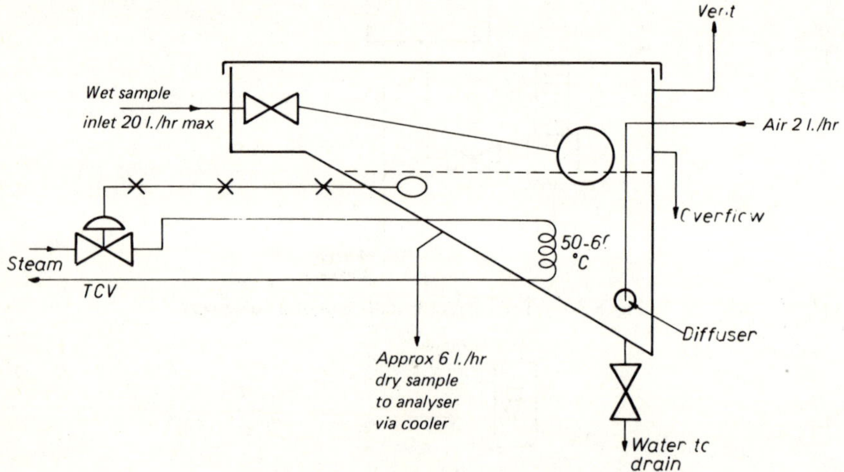

Fig. 8.32 — Home water-stripping tank.

The Panametrics demister (Fig. 8.33 (c)) consists of a number of glass balls contained inside a short length of stainless steel tube. The blowdown ratio is typically between 3:1 and 6:1. If the operating pressure is high, the balls should be of borosilicate glass. Selas also make demisters.

A vapour-in-liquid knock-out pot consists of a suitable vapour disengaging surface of large area, maintained by a suitable level controller which may consist of a valve in the inlet line. Heating, agitating and stripping with air all help to increase the rate of removal of the vapour from the liquid.

A liquid-in-vapour knock-out pot consists of a separator, from which the bottoms liquid is emptied, preferably automatically, by a ball-float and valve [Fig. 8.33 (b)] .

The separator is often centrifugal, with a tangenitial entry. The Hartmann and Braun knock-out pot is illustrated in Fig. 8.33 (d). The coolant can be cold brine from a cooler.

In order to remove the liquid phase from the gas to be analysed, it is sometimes necessary to use a stripper column [Fig. 8.33 (e)] . The packing improves the contact of the air with the liquid sample, so that the amount of air required to strip out the gas is minimized. Such a stripper could be used for a hydrocarbon-in-condensate analyser. An inert gas such as nitrogen may be preferred.

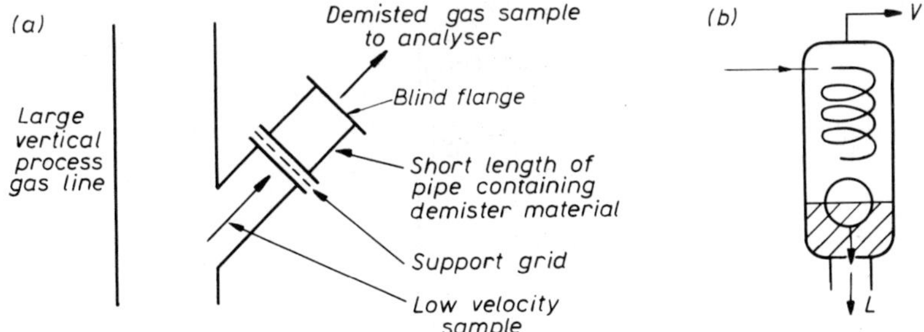

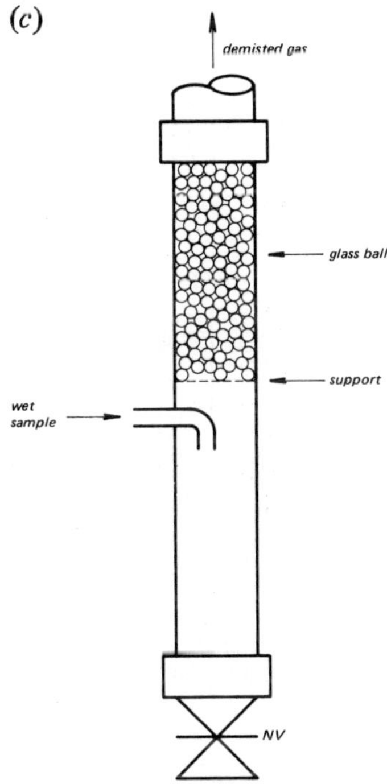

Fig. 8.33 — (a) Use of a demister. (b) Liquid in-vapour knock-out pot. (c) The Panametrics demister.

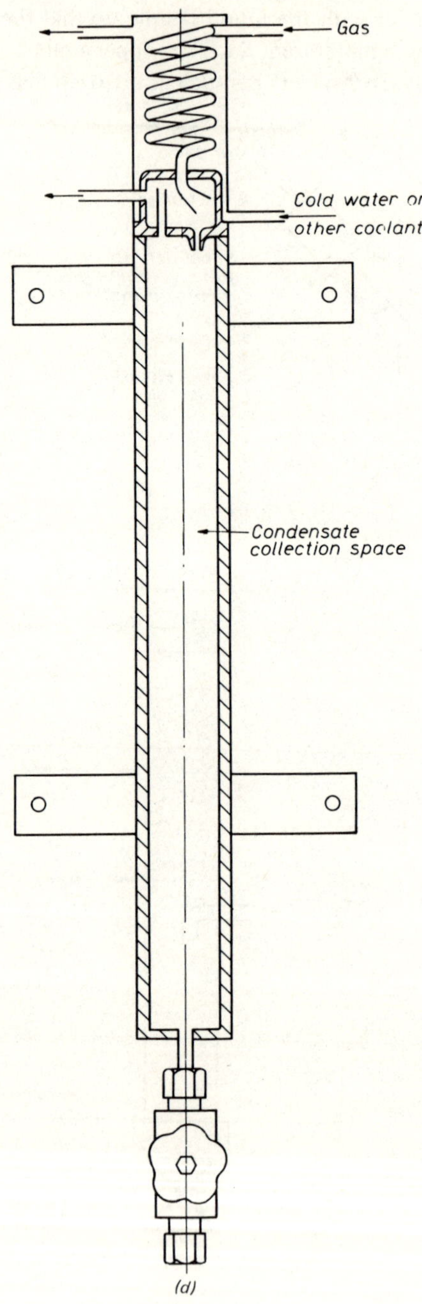

Fig. 8.33 (*d*) – Hartmann and Braun knock-out pot.

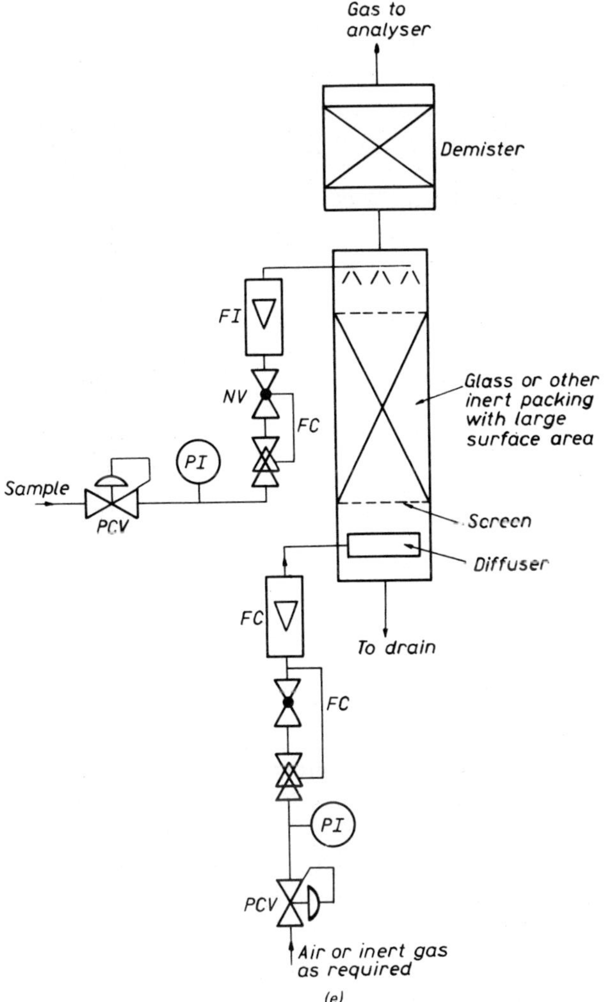

Fig. 8.33 (e) – A stripper column.

The 'Perma Pure' drier (Fig. 8.34) uses the principle of permeation distillation to dry a wet gas sample by passing the sample through a bundle of tubes which are selectively permeable to water vapour. It is claimed that the drier can handle stack-gas samples containing up to 50% v/v water vapour and can reduce the water content by a factor of 400.

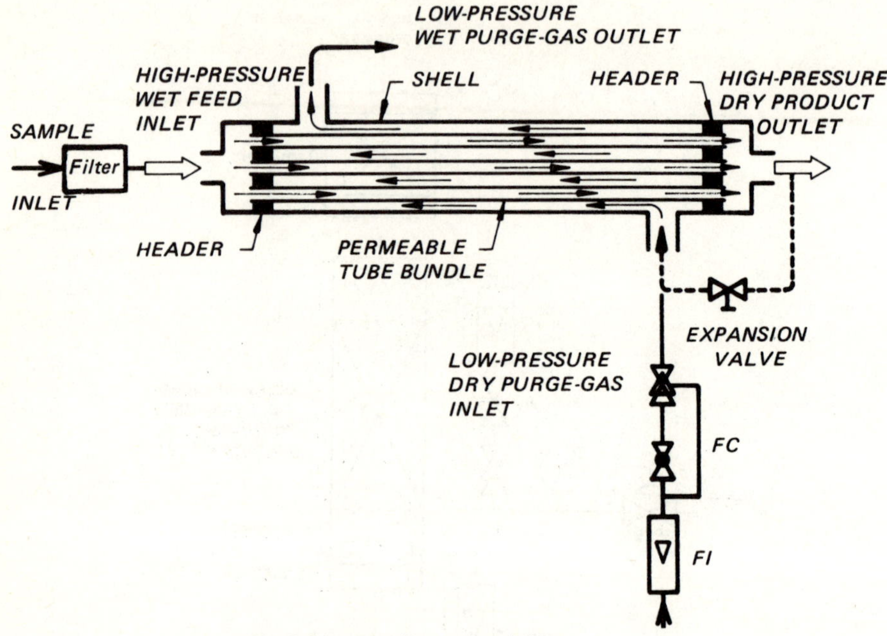

Fig. 8.34 — The 'Perma Pure' drier.

8.5.4 Pressure and flow regulation

Inlet pressure regulation is provided to limit the sample pressure to a safe level for the analyser components and to provide some form of flow control (Figs. 8.35 and 8.36). Whenever a blind pressure-regulating valve is fitted, a pressure gauge should also be fitted. To obtain improved flow control a flow controller should be fitted (Fig. 8.37). To maintain an adequately constant analyser-cell pressure, despite atmospheric pressure variations, an absolute back pressure regulator (Fig. 8.38) should be fitted.

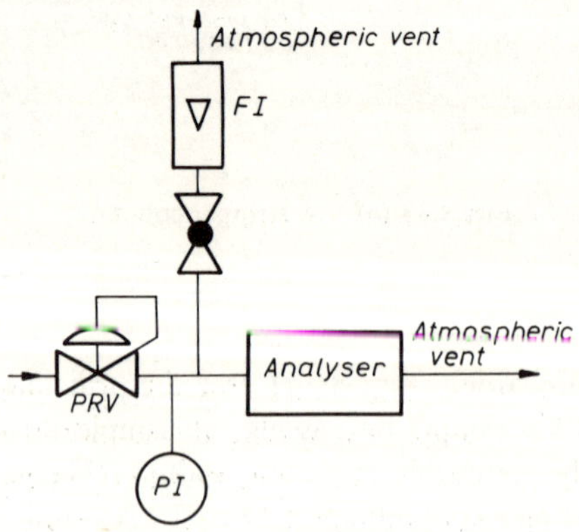

Fig. 8.35 — Use of inlet pressure regulation.

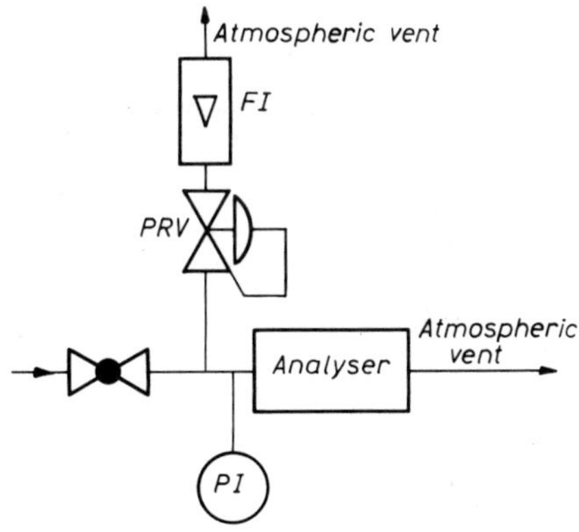

Fig. 8.36 — Use of vent back-pressure regulation.

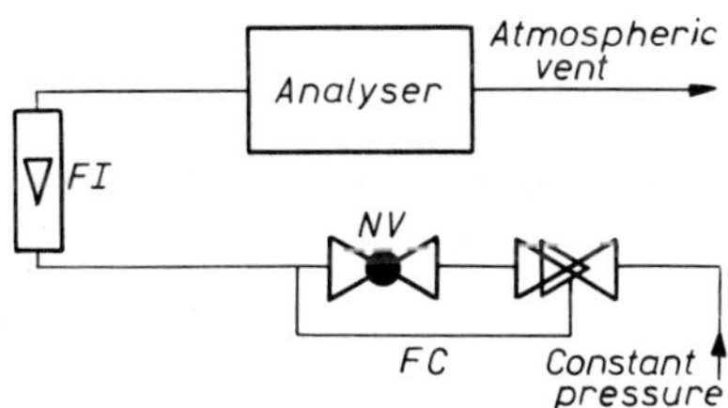

Fig. 8.37   Use of a flow controller.

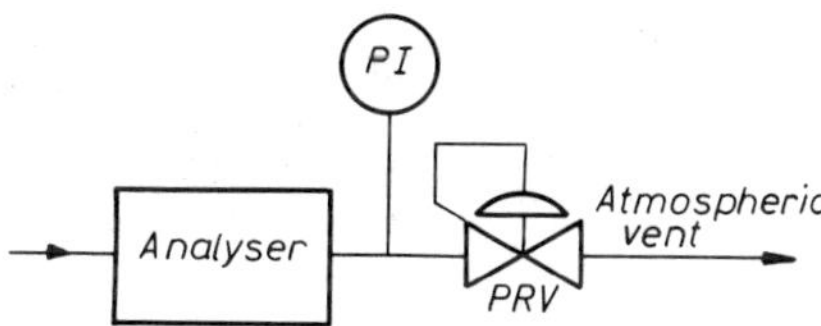

Fig. 8.38 — Use of an absolute back-pressure regulator.

To prevent flashing of a liquid sample, the sample should be maintained at a pressure which is at least 1.5 bar above the vapour pressure at the highest working temperature. For this purpose, a back-pressure regulator may be required (Fig. 8.39).

If the gas-sample pressure is unacceptably large or variable and the pressure has to be reduced or regulated downstream to suit the particular analyser requirements, the pressure-control valve should be located as near as possible to the sample take-off point or fast loop in order to minimize the upstream volume and the resultant time lag (Fig. 8.40).

If the sample is a liquid and the analyser requires a vapour-phase sample, vaporization may be done in the sample conditioner in a heated vaporizing

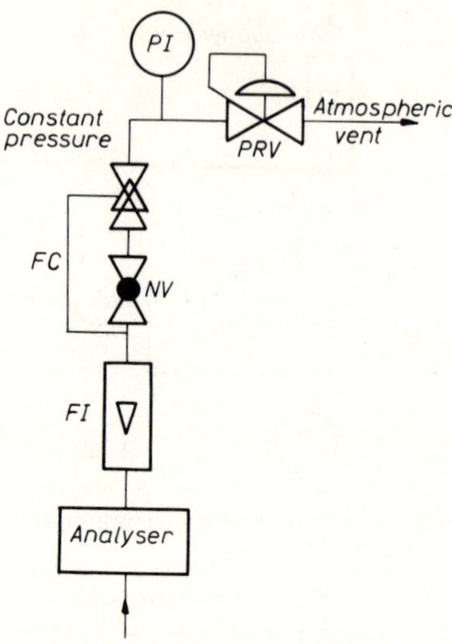

Fig. 8.39 – Use of a back-pressure regulator to prevent flashing.

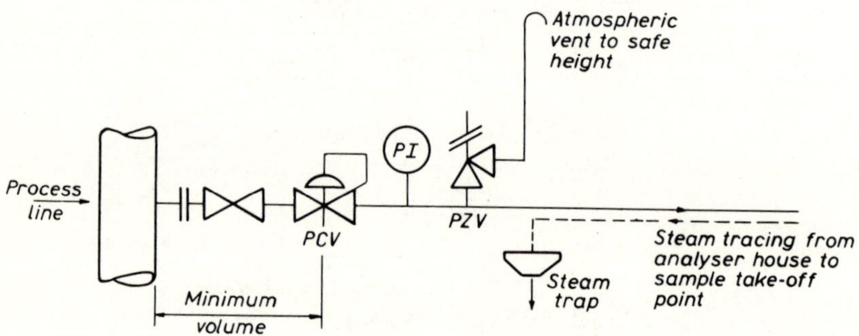

Fig. 8.40 – Sample-pressure reduction at the take-off point.

regulator. If the vaporizer is close to the process line, the sample line will have to be heat-traced to prevent condensation, and such a system should not be used for unsaturated hydrocarbons which can polymerize easily in a sample line.

When a liquid sample is vaporized close to the sample take-off point by a vaporizing reducer, it is essential to minimize the volume upstream of the reducer. If a sample probe, an isolating gate-valve and a Y-type strainer are installed, the volume and resulting lag can be unacceptably high. This lag can be reduced by connecting the strainer drain to a suitable return point. As a result, the slow flow volume element between sample take-off point and analyser is reduced from V_1 to V_2 (see Fig. 8.41). The liquid feed needed to give a vapour flow-rate of 2.1 l./min from the vaporizer is given in Table 8.9.

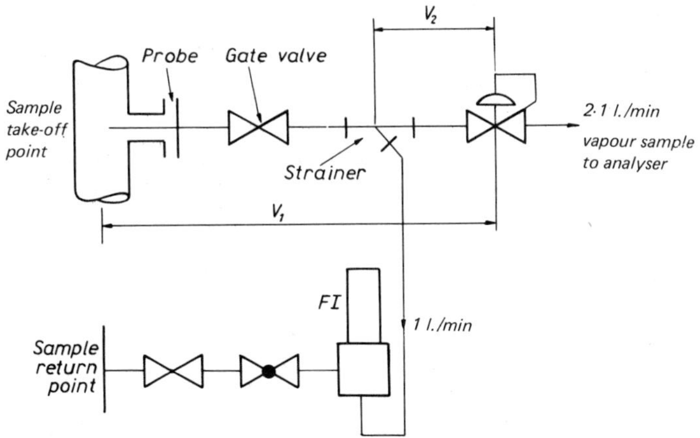

Fig. 8.41 – Local fast loop system.

Table 8.9

Typical vaporizer performance.

	Propane	Butane
Vapour volume flow-rate (ml/min)	2100	2100
Molecular weight	44	58
Density at 1.5 bar ga (kg/m³)	4.9	5.5
Mass flow-rate (mg/min)	10300	12200
Liquid density (mg/ml)	450	510
Volume flow-rate (ml/min)	23	24
Return volume flow-rate (ml/min)	1000	1000
Probe volume flow-rate (ml/min)	1023	1024

The volume V_1 of a $\frac{1}{4}$ in. line system could be about 50 ml and the resultant lag for a liquid propane feed would be about 2 min. The volume V_2 for the modified system could be about 14 ml and the resultant lag only

$$\frac{V_2}{23 \text{ ml/min}} + \frac{V_1 - V_2}{1000 \text{ ml/min}}$$

$$= \frac{14}{23} + \frac{50 - 14}{1000} \simeq 0.6 \text{ min}$$

If this strainer by-pass system is not possible and the lag time must be about 1 min, extra precautions must be taken to minimize the upstream volume. It is sometimes possible to omit the strainer, which could have a volume of about 30 ml, and to reduce the bore of the probe. Thin-walled small-bore pipe should not be used for the sample probe if there is any danger of vibration fatigue causing fracture and consequently damage of expensive equipment downstream of the probe.

If the fast loop is external to the analyser house, then the line connecting the fast loop with the vaporizer must be short and, for example, preferably of $\frac{1}{8}$ in. tubing. The volume of 2 m of $\frac{1}{8}$ in. o.d. tubing is 6.28 ml. For a sample flow-rate of about 25 ml/min, this will produce a lag of about 15 sec.

A valve orifice size may be described according to its sizing constant C_v. For liquids

$$C_v \cong 0.037 F \sqrt{\frac{\rho}{\Delta p}}$$

$$\cong 1.16 F \sqrt{\frac{G}{\Delta p}}$$

where F is the flow-rate in m^3/hr, ρ is the density of the liquid in kg/m^3, Δp is the pressure drop across the valve (in bar) and G is the specific gravity under flow conditions.

For gases

$$C_v \cong \frac{F\sqrt{g\,T}}{295\sqrt{\Delta p\,(p_i + p_o)}} \qquad \text{if } p_o \geqslant \frac{p_i}{2}$$

or

$$C_v \cong \frac{F\sqrt{g\,T}}{257 p_i} \qquad \text{if } p_o < \frac{p_i}{2}$$

where F is the flow-rate in m^3/hr, p_i and p_o are the inlet and outlet pressures in bar abs, T is the temperature of the gas in K and g is the specific gravity of the gas at 273 K and 1.01325 bar.

The C_v of a pressure-reducing valve is related to the diameter d_o (mm), of an equivalent sharp-edged orifice by the equation:

$$C_v \cong 0.027 d_o^2.$$

If sufficient pressure drop is available, flow can be controlled by maintaining an adequate pressure drop across a needle-valve. If a large pressure drop is not acceptable, a lower pressure drop can be used in conjunction with a differential pressure-control valve and a needle-valve. For constant upstream pressure, the needle-valve must be located downstream of the differential pressure-control valve. For constant downstream pressure the needle-valve must be located upstream of the differential pressure-control valve.

When a liquid sample is flowing through a valve with a high pressure drop across it, some flashing may occur. If this interferes with flow measurement, the valve should be mounted downstream of the variable area flowmeter, not upstream. This obviously applies to flow controllers using the needle-valve to produce the required differential.

If flow control is required when flashing can occur, the liquid should be cooled sufficiently or a metering pump should be used. A small reciprocating piston pump is often suitable. The check-valves are part of the pump assembly. The flow-rate can be measured by timing the filling of a burette. Flow control may be upstream or downstream of the analyser, depending on whether the inlet or the outlet pressure is controlled.

The small variable-area flowmeters used for analysers are usually graduated linearly from 10 to 100% and are calibrated with water or air. For liquids other than water,

$$F_{H_2O} \cong F_L \sqrt{\left(\frac{7 \times G_L}{8 - G_L}\right)\left(\frac{7}{\rho_f - 1}\right)} \ .$$

For a sample flow-rate of 0.25 m^3/hr of LPG liquid with $G_L = 0.45$ and a stainless-steel float of density (ρ_f) $\cong$ 8 g/cm^3

$$F_{H_2O} = 0.25 \sqrt{\frac{3.15}{7.55}} \cong 0.16 \text{ m}^3/\text{hr} \ .$$

For gases other than air, with specific gravity g at 273 K and 1.01325 bar,

$$F_{\text{air (0°C, 1.01325 bar)}} = F_G \left[\left(\frac{1.033}{p} \frac{T}{273} g\right)\left(\frac{8}{\rho_f}\right)\right]^{1/2} \ .$$

For a sample flow rate of 0.7 m^3/hr of LPG gas with $g = 1.07$ at 0°C and 1.01325 bar, $p = 1.5$ bar abs, $T = 288$ K and $\rho_f \cong 8$,

$$F_{\text{air (0°C, 1.01325 bar)}} \cong 0.7\sqrt{0.78} \cong 0.5 \text{ m}^3/\text{hr} \ .$$

The meter range might be 0.1–1.0 m^3 of air per hr (at 0°C, 1.01325 bar).

If the sample line to the analyser breaks, flow is excessive, and it can be arranged that the flow switch (FSH) latches off the analyser sample and may only be reapplied when the line has been repaired and the push-button XS has been depressed (see Fig. 8.42).

Alternatively an excess-flow check-valve could be used. This could typically be set at 1.5–2 times the normal rate of flow. The valve closes at the preset excess flow-rate and cannot open again until the back-pressure is sufficient, e.g. the downstream line break has been repaired. A suitable excess-flow check-valve for analysers is the Circle Seal model P–526.

The valve should be located with the fast-loop by-pass sample conditioning system outside the analyser house, so that leakage within the analyser house is limited (Fig. 8.43).

The sample may flow into an atmospheric header tank to produce a constant head of liquid (Fig. 8.44). For liquids, an accumulator such as the Green–Mercier accumulator may be used to absorb the pressure variations.

The use of a volume chamber partially filled with a suitable gas is not successful, since the gas can be gradually pumped out of the chamber by absorption into the liquid. The solution to this problem is to use a barrier between the liquid and the gas, such as a piston or the bag of the Green–Mercier accumulator.

To smooth out large variations in gas-flow through an analyser, caused by a positive displacement pump, it is necessary to include a restrictor and/or volume chamber (Fig. 8.45).

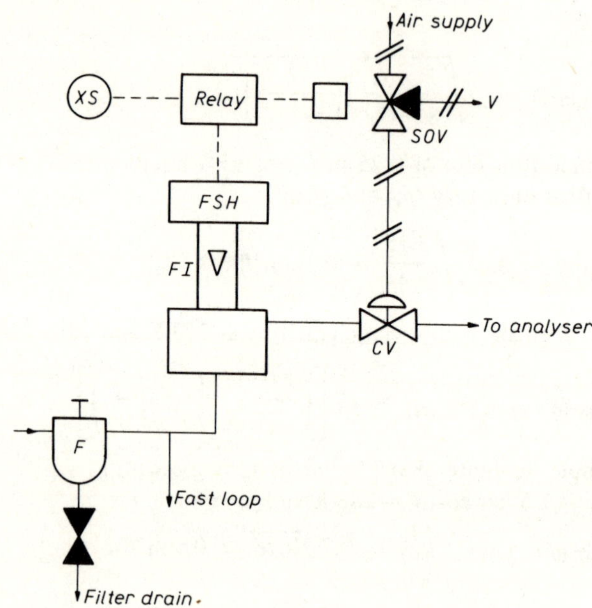

Fig. 8.42 – Use of flow switch to close line to analyser when the line is broken or leaking.

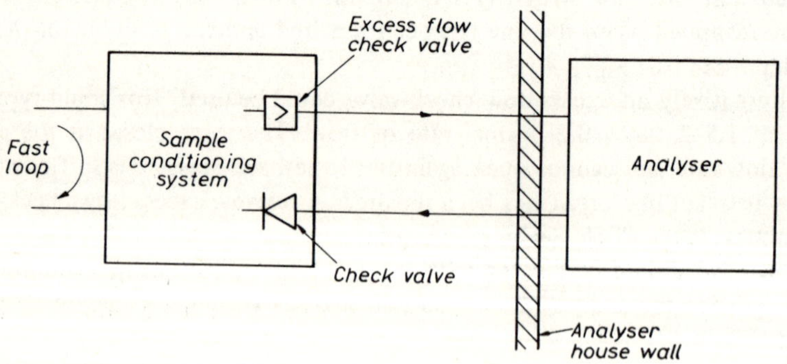

Fig. 8.43 – Use of excess-flow check-valve.

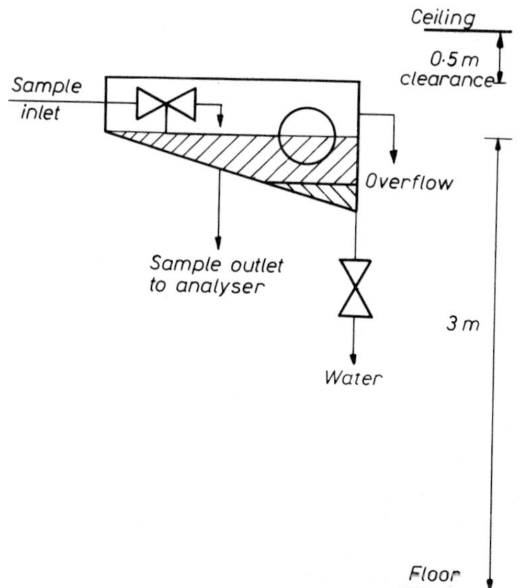

Fig. 8.44 — Atmospheric header tank.

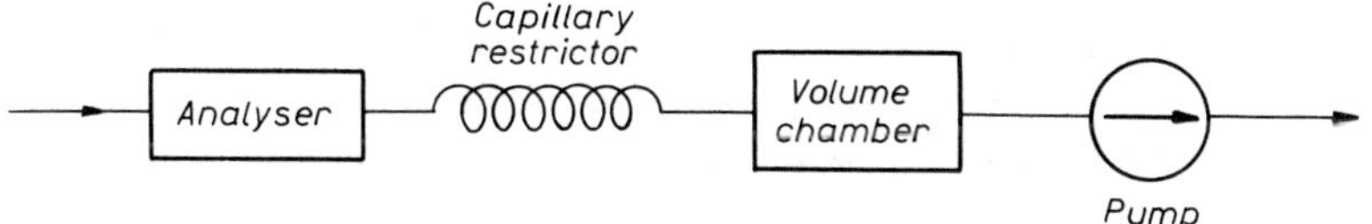

(a)

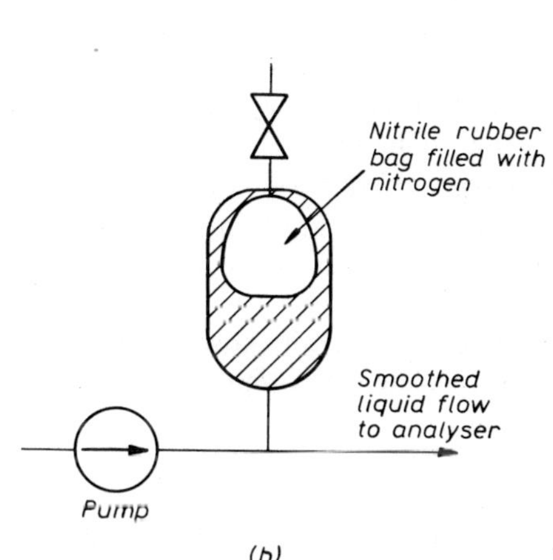

(b)

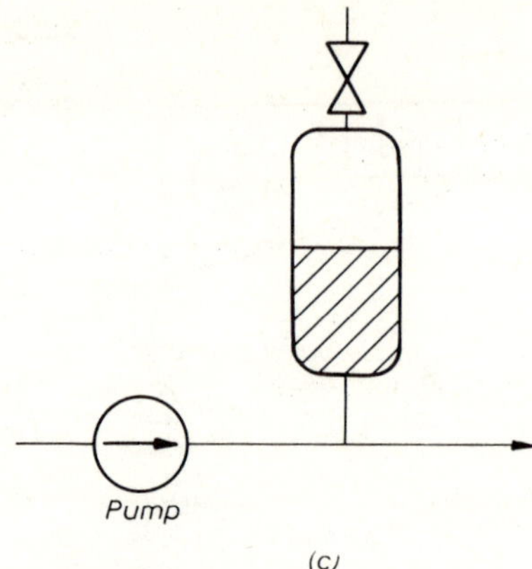

Fig. 8.45 – Methods of smoothing out large gas flow variations.

8.5.5 Control of temperature

Three types of liquid–liquid heat exchangers are commonly used:

    (i) The submerged helical coil.
    (ii) The straight tube-in-tube or pipe-in-pipe.
    (iii) The tube-in-tube helical-coil.

Straight pipe-in-pipe or coiled tube-in-tube exchangers may easily be made on site by using heat-exchanger T-junctions such as the Swagelok products. The design of heat-exchangers for water and steam samples is described in ASTM D1192-70.

Finned air-coolers are also used, especially for samples much hotter than the surrounding atmosphere. The heat may be removed by convection, by a natural draught or by a forced draught due to a fan or blower.

If the cooler is not required to prevent flashing or to cool the sample near to the take-off point, it is better to locate it after the slipstream, where the flow is less and, thus, the size of the cooler may be smaller (Fig. 8.46).

Temperature control minimizes the temperature error of the instrument or the amount of compensation required. The temperature of either the coolant or of the sample by-passing the cooler may be regulated (Fig. 8.47). Well-regulated temperature control can be difficult owing to the lags which normally exist.

Only sufficient cooling water to remove the excess heat need flow, so the cooling water may be regulated according to the temperature of the water discharged from the cooler.

The temperature of the outlet water from the cooler must not be above the maximum allowable value specified for the process plant, unless it is going to an open local drain. In practice, the quantity discharged is so small that it can be neglected.

Sometimes there is an extreme shortage of sufficiently clean cooling water.

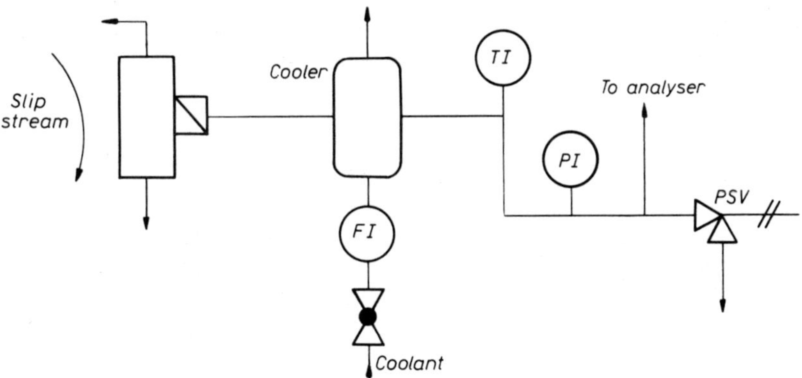

Fig. 8.46 – Location of cooler after the slipstream.

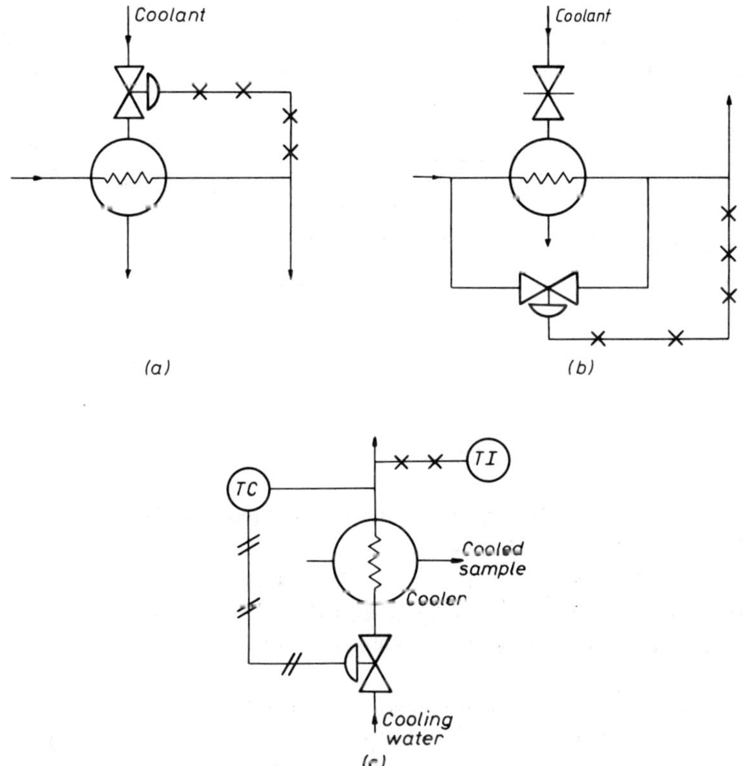

Fig. 8.47 – Various methods used for controlling sample cooling.

It is better then to recirculate the clean water and to cool it with the dirtier plant-cooling water, which should never be used for direct analyser-sample cooling (Fig. 8.48), and must, of course, be cold enough.

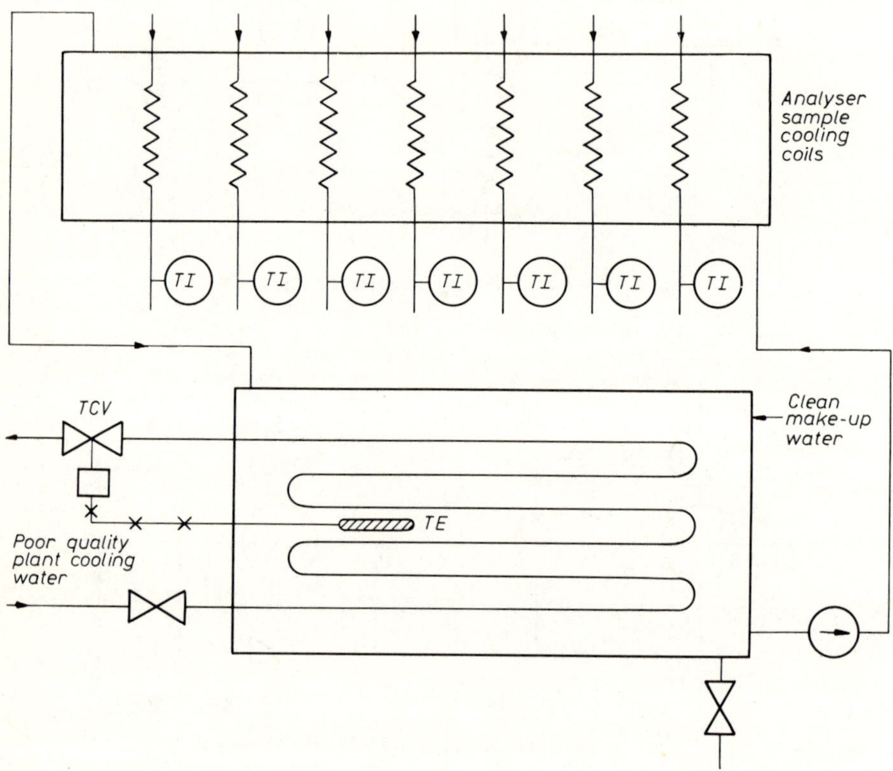

Fig. 8.48 – Closed clean cooling-water system

The temperature controller is not often required since the supply of plant cooling-water is usually not limited.

On occasion it is advantageous to use a simple mechanical refrigerator. This avoids the problems of analyser cooling-water freezing in winter. Slow moving cooling water is always more likely to freeze than faster moving cooling water.

Peltier cooling is only economical if a small amount of heat has to be extracted at a controlled rate. If the polarity of the d.c. electrical supply is reversed, the Peltier cooler will then act as a heater. This is used in the Hallikainen pour-point analyser. It is advisable to check the reliability history of any Peltier cooler before installation.

Air cooling is sometimes sufficient, especially when the temperature difference between the sample and ambient air is large. It is always relatively simple to apply, especially when the draught is natural and not forced.

In an LPG or NGL plant, or where there is sufficient propane, dry purified liquid propane may be used to cool a sample (Fig. 8.50). The temperature of the expanded gas depends upon the final gas pressure. The amount of cooling depends on this pressure, the latent heat of vaporization of the propane, and the rate of flow of propane into the flash cooler. The final temperature can be below $0°C$, so the propane should be bone dry.

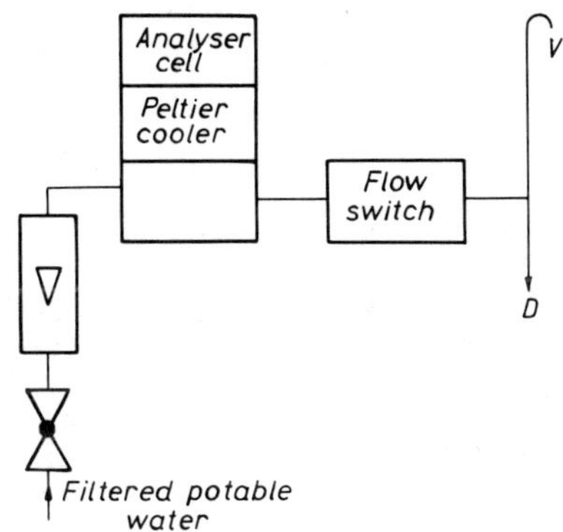

Fig. 8.49 – Use of a Peltier cooler.

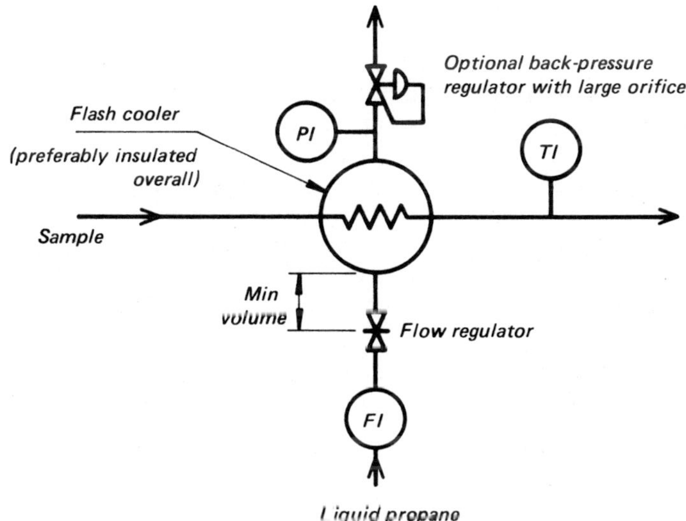

Fig. 8.50 – A liquid propane flash cooler.

Automatic temperature control is possible but is also seldom used in practice. When it is used, the by-pass method is preferred since it is not easy to make a satisfactory control valve for propane for this application.

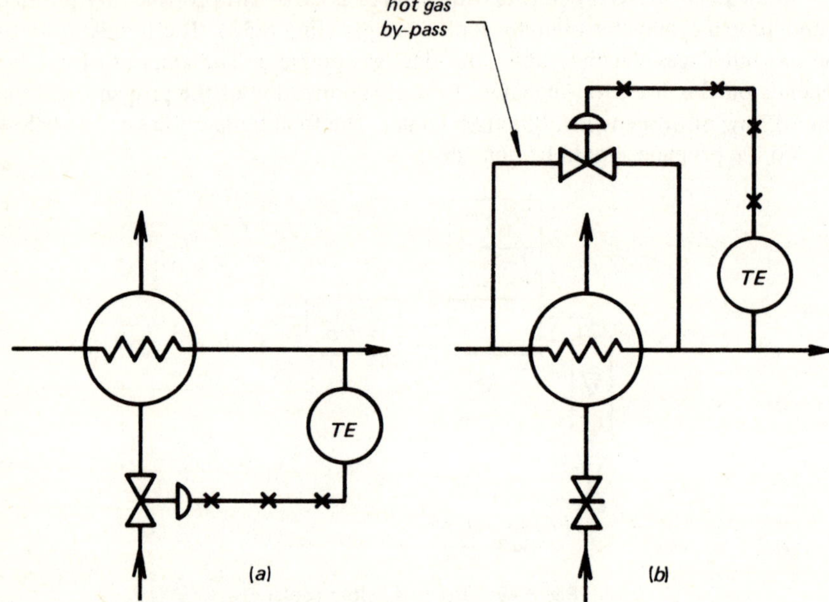

Fig. 8.51 – Automatic temperature control, (*a*) by regulation of the coolant, and (*b*) by hot gas by-pass.

Suitable stainless-steel coolers are the Graham Manufacturing Heliflow and the Alfa Laval Lamella (Fig. 8.52). The Heliflow cooler consists of several parallel tubes wound into a multi-layer spiral coil.

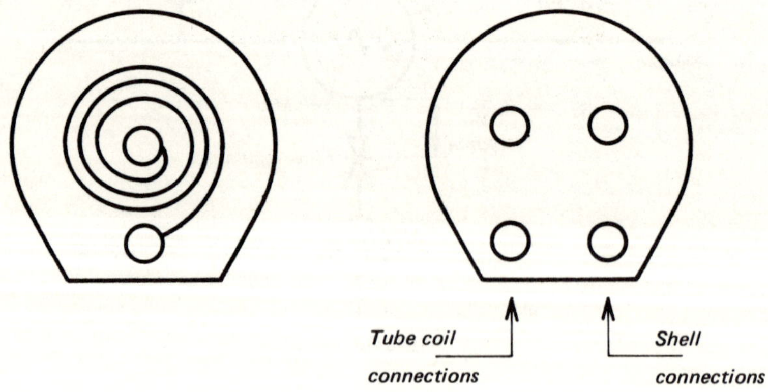

Fig. 8.52 – Cooler designs.

This method of cooling is particularly useful for Reid vapour pressure analyser sample systems in very hot climates where sufficiently clean water is seldom cool enough and is in short supply.

If the cooler is used only to make the measurement more accurate or to maintain a low vapour pressure to minimize flashing, then cooling-water failure is not too serious. However, if the cooler is used to protect the analyser from overheating, and hot sample could damage the analyser on cooling-water failure, or if any hazard could arise from failure, the sample must be capable of being isolated (Fig. 8.53) if failure occurs. A flow switch (Gems or Brooks) is better than a temperature switch, which acts more slowly.

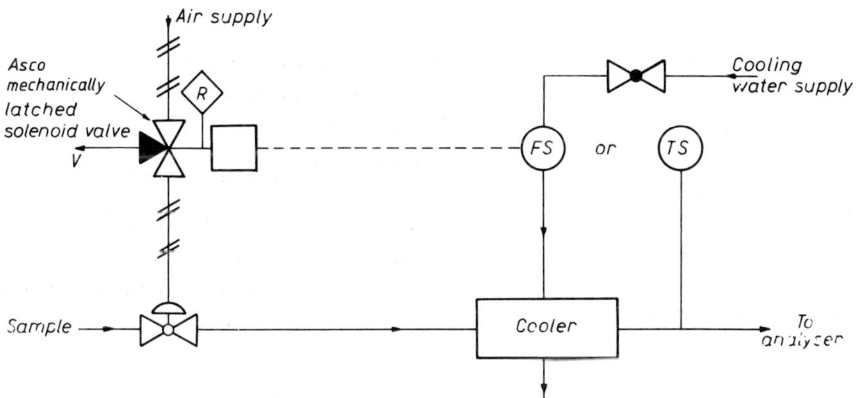

Fig. 8.53 — Methods of isolating the sample on cooling water failure.

When selecting coolers to prevent flashing of liquid samples, first check the values of the pressure and temperature of the process fluid and use tables to find the saturation pressure at the working temperature. The pressure can be reduced to this pressure (but not below it) without flashing. This may then be followed by reduction of the sample temperature. If the sample pressure is near to the saturation pressure at the working temperature, the sample should be pre-cooled before pressure reduction, with final cooling to the sample temperature required for the analyser. Either a lower temperature reducer or a lower pressure sample cooler may be cheaper. The analyser element need not necessarily be at atmospheric pressure and ambient temperature, provided that these parameters and their variations are acceptable.

When the pressure of a vapour sample is reduced, the temperature of the sample is also reduced. In order to maintain the vapour phase, it must be ensured that the cooling is not sufficient to cause condensation, i.e. that the resultant sample temperature is above that of the dew point.

As an example, consider the reduction of pressure of a light hydrocarbon gas from 6.4 bar abs at 17°C (290 K) to 2.5 bar abs.

For such a gas the ratio of the specific heats at constant pressure and constant volume respectively, $C_p/C_v = \gamma$, will be about 1.25.

If it is assumed that the gas is a perfect gas and that the process is reversible and adiabatic, then

$$T_2 = T_1 \left(\frac{P_2}{P_1}\right)^{(\gamma - 1)/\gamma}$$

$$= 290 \left(\frac{2.5}{6.4}\right)^{0.2}$$

$$= 240 \text{ K} = -33°C$$

Assume a critical pressure of 130 bar abs and a hydrocarbon dew point of 4.4°C.

The equilibrium ratio (K_{CO_2}) of CO_2 for these conditions lies between that of methane (C1) and ethane (C2) and is given by

$$K_{CO_2} = \sqrt{K_{C1} K_{C2}} = \sqrt{6.01 \times 1.1} = \sqrt{6.6} = 2.6$$

The equilibrium ratio is defined as the ratio of the mole fraction of the component in the gas phase to that in liquid phase, and tables are available in the literature [5], to which the reader is referred for details of the calculations. For a typical gas mixture the following values will be obtained.

Component	CO_2	N_2	C1	C2	C3	IC4	NC4	IC5	NC5	NC6	Total
Mole fraction in gas phase, Y_i	0.047	0.084	0.72	0.07	0.45	0.006	0.016	0.004	0.005	0.0019	$\cong$ 0.99
K_i	2.6	16	6	1.1	0.3	0.12	0.09	0.038	0.031	0.011	
Mole fraction in liquid phase, $X_i = Y_i/K_i$	0.018	0.005	0.12	0.16	0.15	0.05	0.18	0.11	0.16	0.17	$\simeq$ 1.02

ΣY_i and ΣX_i should both be 1.00 .

The approximations are sufficiently accurate, so we may assume that the hydrocarbon dew point is 4.4°C.

The sample was taken downstream of a dehydrator, so the moisture dew point was much less than the hydrocarbon dew point.

If the cooling were adiabatic, much condensation would be expected, and this would result in a sample error, but in practice, the sample line and valve are not insulated and sufficient heat is gained from the surrounding air for the cooling of the sample to be much less.

If a temperature reduction of 1.6°C for 1 bar pressure reduction is assumed, the temperature would be

$$T_1 - 1.6\,(P_2 - P_1) = 17 - 1.6 \times 3.9 = 10.75°C \ .$$

This is much higher than the expected dew point, so there should be no condensation of the sample.

If the actual sample temperature after pressure reduction is near to or below the dew point temperature, heat tracing is necessary.

If the amount of expansion cooling is large, part of the line upstream of the valve used for pressure reduction must be heat traced as well as the valve itself. These components should then be thermally insulated. Raychem 'Chemelex' is a useful heat-tracing material, since it is temperature self-limiting and does not require a temperature switch.

In order to prevent vaporization of a liquid sample, the sample temperature must remain below that of the bubble-point at the sample pressure. The bubble-point temperature of the sample can be determined by estimating the temperature at which

$$\Sigma\, Y_i \;=\; \Sigma\,(X_i\,K_i) \;=\; 1.00$$

The procedure is to calculate the molar fractions for the liquid phase (X_i) from the analysis and then to select K_i values for various temperature and pressure conditions that will give

$$\Sigma\, K_i\, X_i \;=\; \Sigma\, Y_i \;=\; 1.00$$

The temperature and pressure corresponding to the K_i values will be those of the bubble-point (or IBP), The following example illustrates a typical calculation (for a convergence pressure [5] of 1000 psia = 22 bar abs).

Component	H_2S	C2	C3	IC4	NC4	Total
Liquid mole fraction X_i	0.001	0.015	0.958	0.019	0.007	1.000
Vapour/liquid ratio K_i for 65°C and 22 bar abs	2.5	2.3	1.02	0.7	0.5	
Vapour mole fraction Y_i $= \Sigma K_i X_i$	0.025	0.0345	0.977	0.0133	0.0035	~1.03
Vapour/liquid ratio K_i for 63°C and 22 bar abs	2.0	2.4	1.0	0.6	0.45	
Vapour mole fraction Y_i $= K_i X_i$	0.002	0.036	0.958	0.011	0.003	$\cong 1.00$

Thus, for this sample, the bubble-point should be at 63°C and 22 bar abs.

Undesirable flashing will occur if the operating temperature margin that can be allowed depends on the normal variations in operating pressure and the expected variation in the bubble-point of the sample.

If the margin is inadequate to allow sample-pressure reduction without flashing, sample cooling is necessary with the sample cooler upstream of the reduction valve.

8.6 STREAM SWITCHING

'Block and bleed' stream switching systems are shown in Fig. 8.54.

In a stream-switching system it is essential to ensure that one sample stream does not interfere with the measurement of the other. The old sample must be swept clear of the system by the new sample as quickly as possible, with the minimum of dead volume. In the block and bleed system shown in Fig. 8.54 (*f*) the first pair of solenoid valves ensures that the samples are always flowing along the sample lines, so that the sample is always up-to-date. The second pair of solenoid valves ensures that all sample volumes are swept through the system as fast as possible with the minimum of dead-space sample diffusion.

However, solenoid valves do frequently leak. Thus, it is better, although more expensive, to replace the solenoid valves by a sample-selection valve such as a slide valve, which is operated by two solenoid valves which are not so critical (in terms of leaking) because they are in the air lines. As alternatives to the system shown in Fig. 8.55, a single 3-port solenoid valve and spring-biased slide valve may be used, or the two 3-port solenoid valves may be replaced by a 5-port solenoid valve. Air-operated ball valves give tight shut-off and thus have advantages for high-pressure liquids.

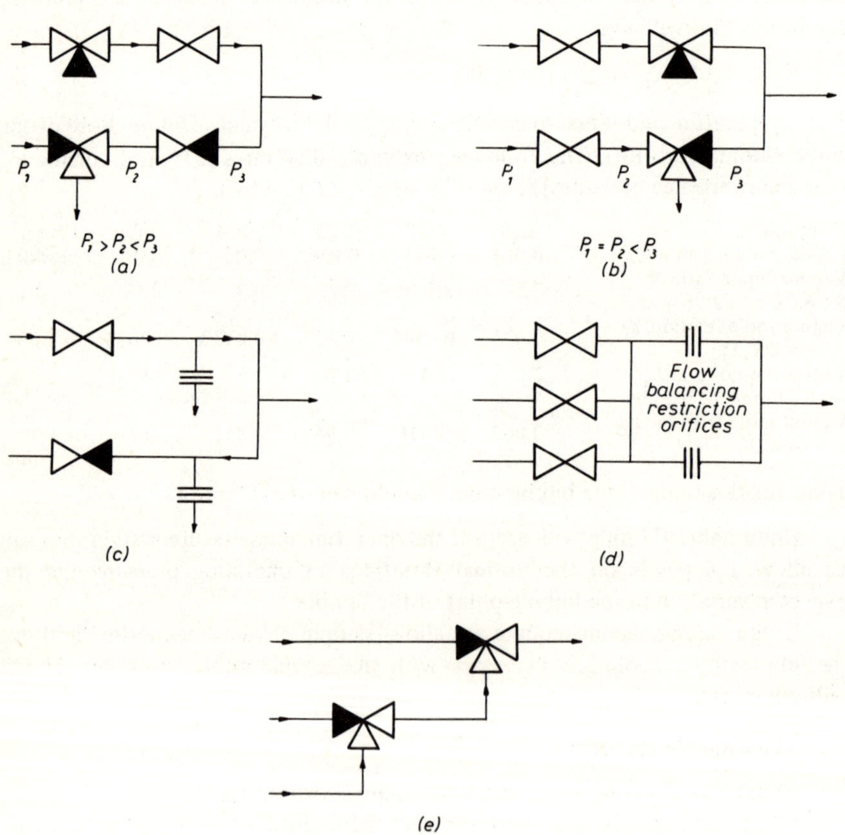

Fig. 8.54 – Block and bleed stream-switching systems.

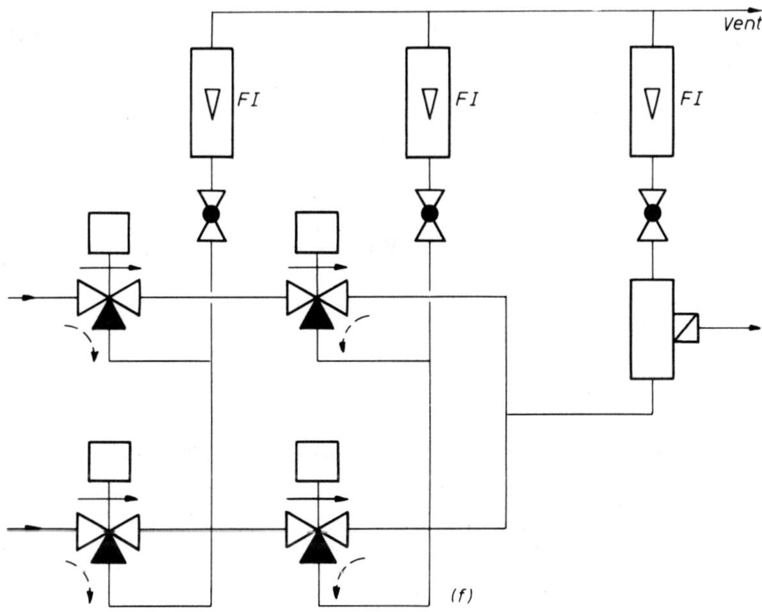

Fig. 8.54(*f*).

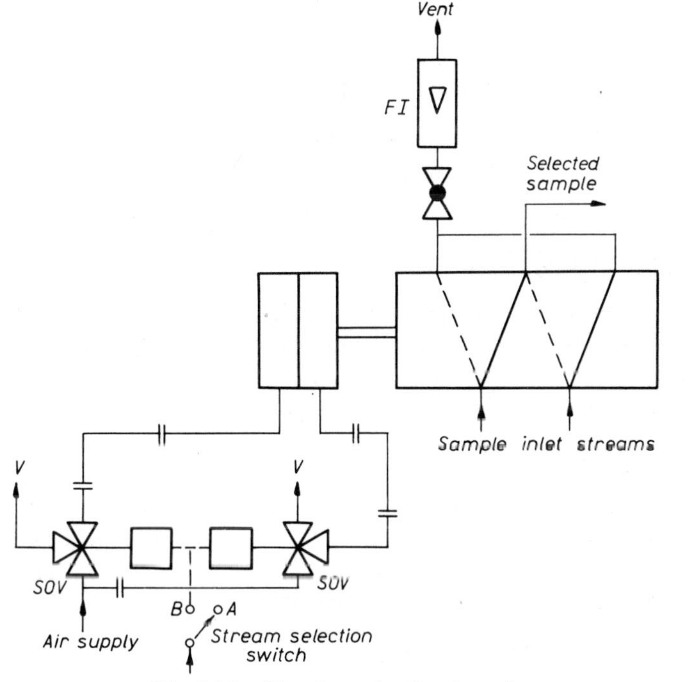

Fig. 8.55 – Use of sample-selection valve.

With the two-stream flow-switching system shown in Fig. 8.56, the switching valves are in the vent lines and not in the analyser sample line.

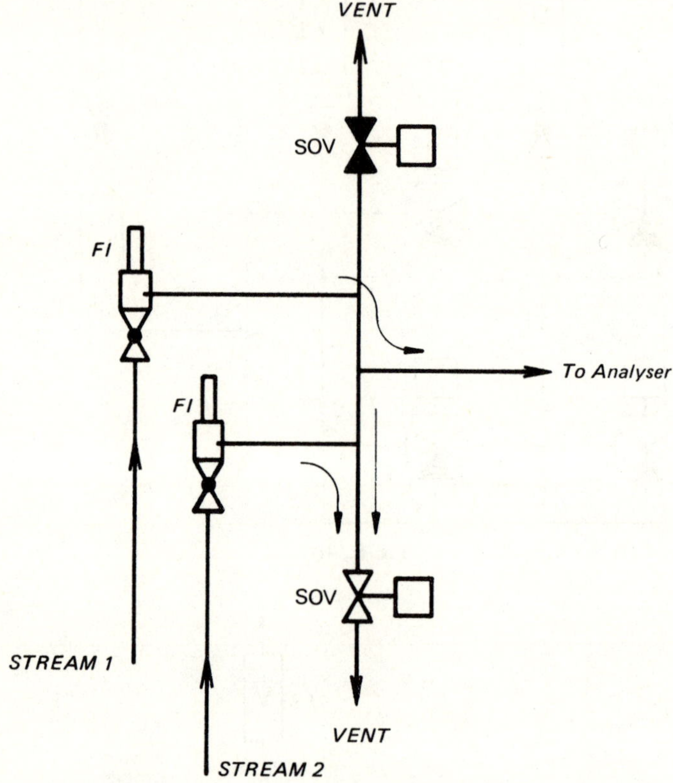

Fig. 8.56 — A two-stream (external) flow-switching system.

8.7 SAMPLE PUMPS

A pump is normally only required to raise the pressure head of a liquid stream at a particular rate of flow. The total differential pressure head produced by a pump is equal to the discharge head, minus the suction head. If suction lift is required then the suction head is considered to be negative:

$$H = H_\mathrm{d} - H_\mathrm{s}$$

To these static pressure heads must be added the friction and velocity heads. To the suction discharge heads must be added the pressure of the liquid at the suction and discharge elevations (Fig. 8.57).

The inlet or brake horse power is given by

$$\mathrm{BHP} = \frac{\text{delivered power}}{\text{efficiency}}$$

The delivered power = constant × flow-rate × total head × liquid density

$$1 \text{ m } H_2O \text{ head} \sim 0.1 \text{ bar}$$

$$= \frac{1 \text{ m liquid head}}{\text{S.G. of liquid}}.$$

The friction head depends upon the viscosity of the liquid.

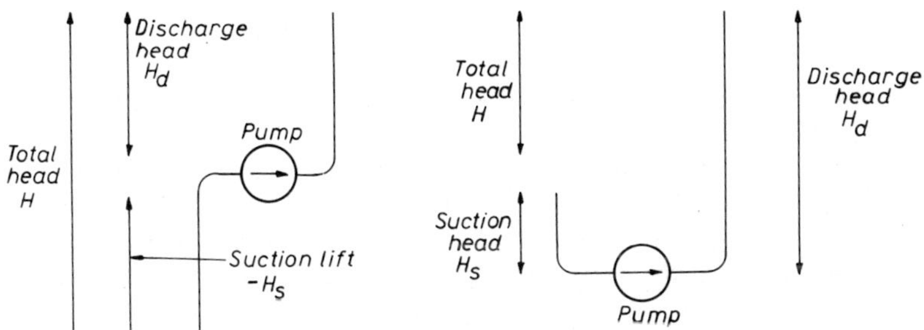

Fig. 8.57 – Illustration of total head, suction head and discharge.

The required Net Positive Suction Head (NPSH) at the pump inlet is the suction head less the vapour pressure at the pump inlet (Fig. 8.58).

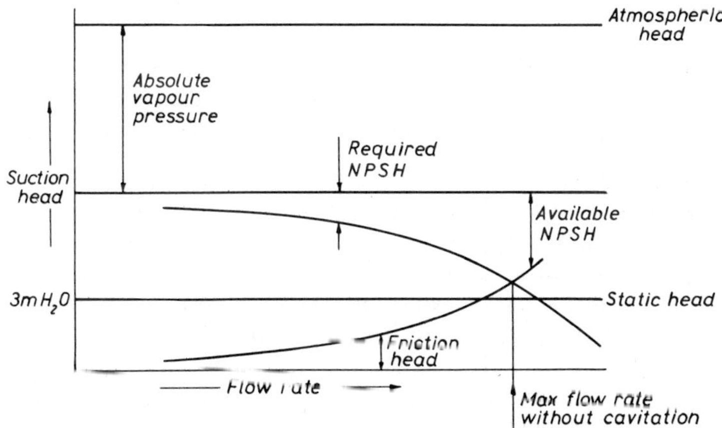

Fig. 8.58 – Variation of suction head with flow-rate.

Cavitation, caused by too high a flow-rate resulting in the downstream pressure falling below the vapour pressure of the liquid, usually causes rapid erosion of the pump where it occurs and must therefore be eliminated.

The theoretical height to which a liquid can be lifted at any particular temperature depends on the difference between the atmospheric pressure and the vapour pressure. In practice it is less. This is illustrated in Fig. 8.59 for water.

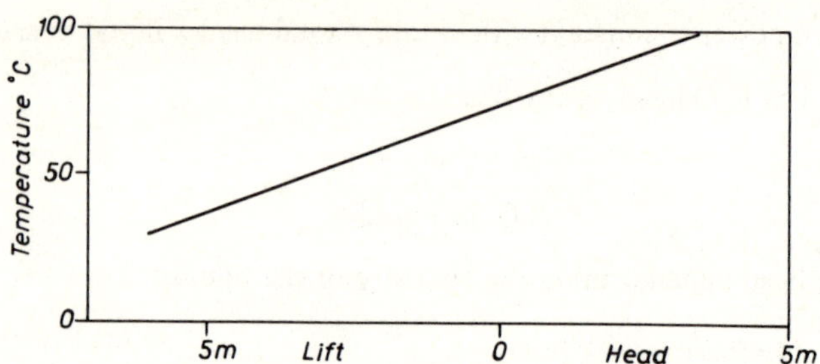

Fig. 8.59 – Theoretical height to which water may be lifted, as a function of temperature.

Apart from providing an increased head, a pump may also be used to provide an accurately predetermined rate of flow of a liquid. Such pumps are known as metering pumps and are used to supply reagents for controlled dosing. They sometimes take the form of a controlled-stroke piston pump where the stroke may be adjusted remotely by a pneumatic or electrical signal.

Sample pumps may be divided into two classes.

(i) *Reciprocating Pumps.* These produce pressure surges which may be reduced if multiple stages are used. The pressure surges can also be reduced by the use of a volume impulse absorber. Capacity control may be obtained by varying the stroke or the rate of stroke of the pump. These are often described as constant volume or metering pumps.

(ii) *Rotary Pumps.* If a rotary pump has a large leakage around the periphery of the rotor, as is the case for centrifugal pumps, then the pump certainly cannot be used as a metering pump but may be suitable as a sample pump.

Capacity control may be obtained by varying the speed of the pump, by varying the rate recirculated or by adjusting a valve in series with it, according to the type of pump and its characteristics.

Pumps may be driven by electric motors with or without reduction gearboxes or by pneumatic motors which may be of the piston or turbine type. In many process-analyser applications, the sample fluid and the surrounding atmosphere can produce an electrical hazard; the drive motor must then be explosion-proof. In petroleum refineries and petrochemical plants, totally enclosed electric motor drives are used with fan cooling where necessary, especially for the larger sizes. NEMA Class B insulation is normally sufficient.

The parts of the pump in contact with the sample fluid must be made leak-tight, preferably for a long time, to minimize maintenance and to maximize analyser availability. The seals should be self-lubricating. Unfortunately seals leak in time, so lantern rings are sometimes used if leaks are important. The seal liquid chosen must be removable from the sample and must not interfere with the analysis. As an example, water is used as seal liquid for a pump handling a flue-gas sample.

Mechanical seals require less maintenance than packed seals, which can work at much higher temperatures. Packed seals tend to leak more, but when they fail they fail less rapidly than mechanical seals.

Bearings may be lubricated internally by the sample fluid, or externally (by a separate lubricant, outside a seal). They may be ball, taper, pin or sleeve bearings. The latter may be carbon or PTFE.

Pumps may be jacketed or traced to ensure that the fluid temperature is maintained within the required temperature range.

Quite often the sample fluid has a low viscosity and thus a low lubricity. Friction in the pump itself must therefore be limited. To minimize pump friction there must be adequate clearances between the sliding surfaces, or rolling surfaces must be used. For this reason gear-type pumps can be used for mid-distillates, but not for LPG for which a turbine-type pump is preferable.

The temperature of the fluid discharged from the pump can increase, owing to heat conduction from the drive, the operation of the pump, and the ambient temperature. The sample fluid can also become heated if the fluid is recycled. Cooling can occur if the sample gas or vapour is expanded.

When the pump discharge is blocked or restricted, the discharge pressure is likely to rise to an unacceptably high level, and should be limited by fitting a pressure relief valve or a back-pressure control valve (Fig. 8.60). Pumps with a large rotor clearance do not require such protection.

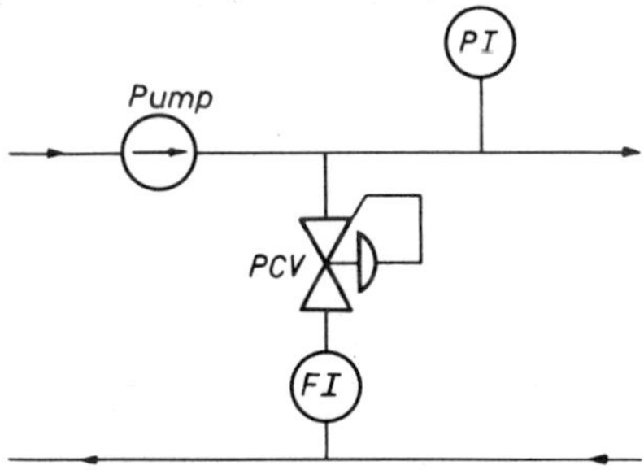

Fig. 8.60 – Pump protection.

If the pump has been designed for liquids, care must be taken that the sample does not contain any vapour, otherwise the internal mechanism of the pump could be damaged. For these reasons, a sufficiently high suction head should always be available and the pump should be self-priming.

If an analyser requires a pump, an outlet pump will introduce less contamination than an inlet pump; an inlet pump will have to withstand a higher working pressure and an outlet pump may have to be able to handle more vapour without cavitation damage.

The pump will probably be located at the inlet to the analyser [Fig. 8.61 (a)]. A back-pressure regulating valve will then perhaps be necessary to limit the maximum pump-case pressure [Fig. 8.61(b)]. The addition of an analyser back-pressure valve increases the pressure in the analyser to a sufficient margin above the vapour pressure of the fluid to prevent vaporization, but the discharge pressure of the pump has also to be increased by the same amount [Fig. 8.61 (c)].

The required pump head can then be reduced by connecting the pump suction to the back-pressure regulator, thus raising both the analyser outlet pressure and the pump suction by the required amount [Fig. 8.61 (d)]. Whereas the pump back-pressure control valve can be used to protect the pump from the development of an excessive case pressure on low flow, perhaps resulting from blocked discharge as in Fig. 8.61 (b), the flow through this valve may also be increased to give a fast sample loop.

Flowmeters are located at high-pressure parts of each loop to minimize vapour in the meter, and reduce the resultant error. If no vapour is normally present, the meters can be relocated at a low pressure point [Fig. 8.61 (e)]. The flow in each loop can be regulated by the addition of needle-valves or perhaps flow controllers.

This system could be used with a vapour-pressure analyser for which a turbine pump should be used. Turbine pumps can withstand up to 20% vapour in the suction line without cavitation damage.

Care must be taken that the proportion of vapour returned to pump suction from the analyser is kept within acceptable limits. Worst-condition calculations must always be made, to ensure that the maximum vapour content is known.

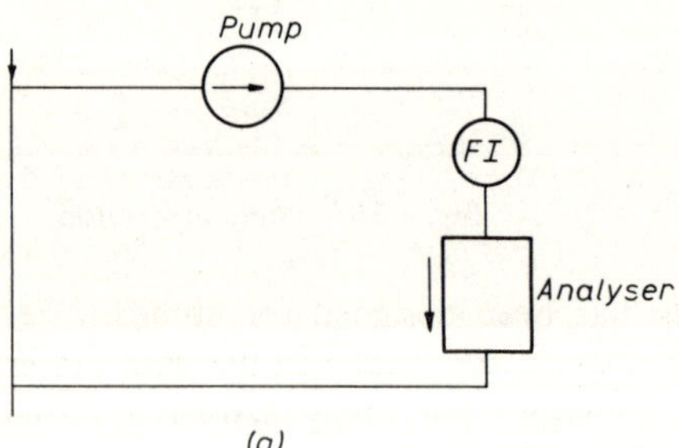

Fig. 8.61 – (a) Pump located at analyser inlet. (b) Use of pressure-control valve with pump. (c) Use of pressure-control valve at analyser discharge. (d) Pump section connected to back–pressure regulator. (e) Fast sample loop.

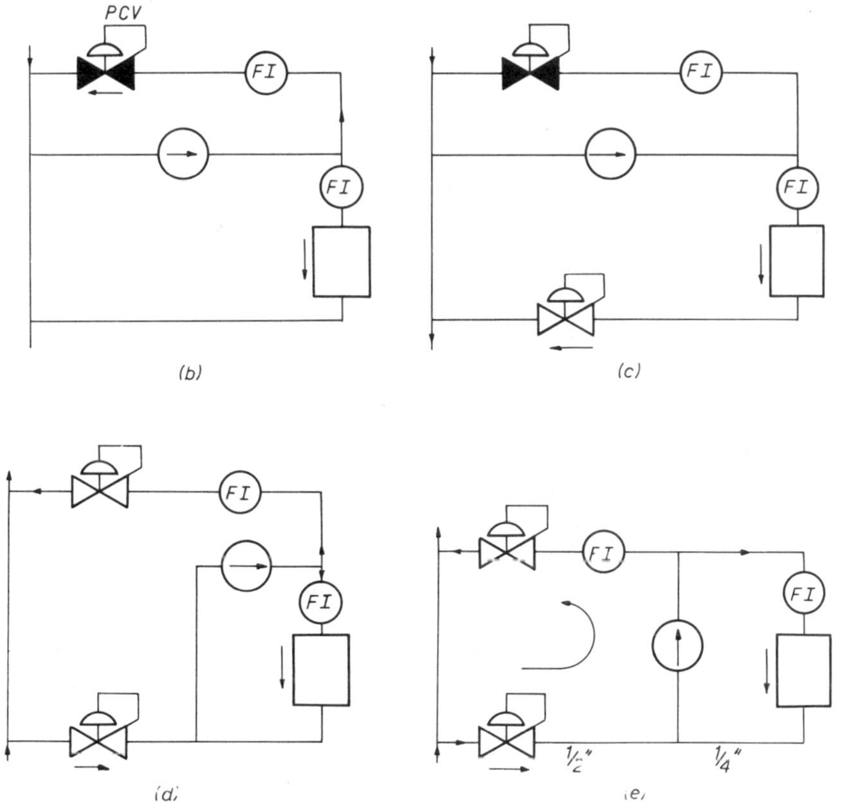

8.8 METHODS OF OBTAINING A SAMPLE AT LOW PRESSURES

The suction pressure should be less than 0.2 bar below atmospheric pressure in order to minimize the amount of air leaking at the joints. A short pipe run giving turbulent flow should be used to prevent phase separation. Figures 8.62 (a) and (b) illustrate two similar systems for regulating and limiting the pressure and pressure drop.

In flue-gas monitoring, the position of the sample point is important. If it is too near to the burner, there is no guarantee that a representative sample will be taken, and at the same time, the high temperatures make construction of the probes difficult.

If the sample point is too far up the flue, the temperature is much lower, but there is a danger that the system is now measuring air leaks into the various joints of the flue rather than the excess of air. As a general principle, the sample point should be as near to the firing point as possible, consistent with adequate flue-gas mixing. For each individual installation, the services available and the degree of accuracy required have to be taken into account.

There are six basic methods of extracting sample gas from the flue:

(*a*)　water aspiration;
(*b*)　air aspiration;
(*c*)　steam ejection at the cold end of the probe;
(*d*)　steam ejection inside the flue;
(*e*)　diaphragm pump;
(*f*)　rotary pump.

(*a*) *Water aspiration.* In the water-aspiration method, a small water-jet pump is employed to produce a low pressure which draws in the flue gas through the sample probe. The flue gas mixes with the water, which is removed in a separator. The gas is then passed into the analyser.

Although this system is simple to instal and operate, it possesses disadvantages, viz.

(i)　Much of the sample line, including the probe, is at a pressure lower than atmospheric. Thus, any leaks in the system will allow air to enter.

(ii)　A large water flow is required to obtain a high sample flow. This water dissolves some of the carbon dioxide in the flue gas, and dissolved oxygen may be released into the sample stream, resulting in a high oxygen reading.

(iii)　The flue gas cools in the sample line before reaching the aspirator, and forms a corrosive condensate. The materials used must therefore be able to withstand this condensate, or water should be injected into the sample system after the probe in order to ensure that the condensate is diluted.

There are compensating advantages, however.

(i)　The gas is washed and dirt or any corrosive constituent is, to a large extent, removed before the gas is passed through the analyser.

(ii)　The gas is cooled and therefore is less likely to cause condensation in the analyser.

Water-operated aspirators are supplied from either a constant-head tank or direct from the mains through a pressure regulator. They can also be used for suction aspiration where the sample should not come into contact with water before analysis. Great care must be taken with this method to prevent leaks in the analyser sampling connections.

Separators are used only in pressure-aspiration systems. They must be mounted at least 1.5 m below the level of the aspirator. The optimum height between the aspirator and the separator is 3–3.5 m. The water separated from the gas sample is passed to the drain through a water seal.

The Hartmann and Braun water-operated gas-jet pump is shown in Fig. 8.24 (*c*).

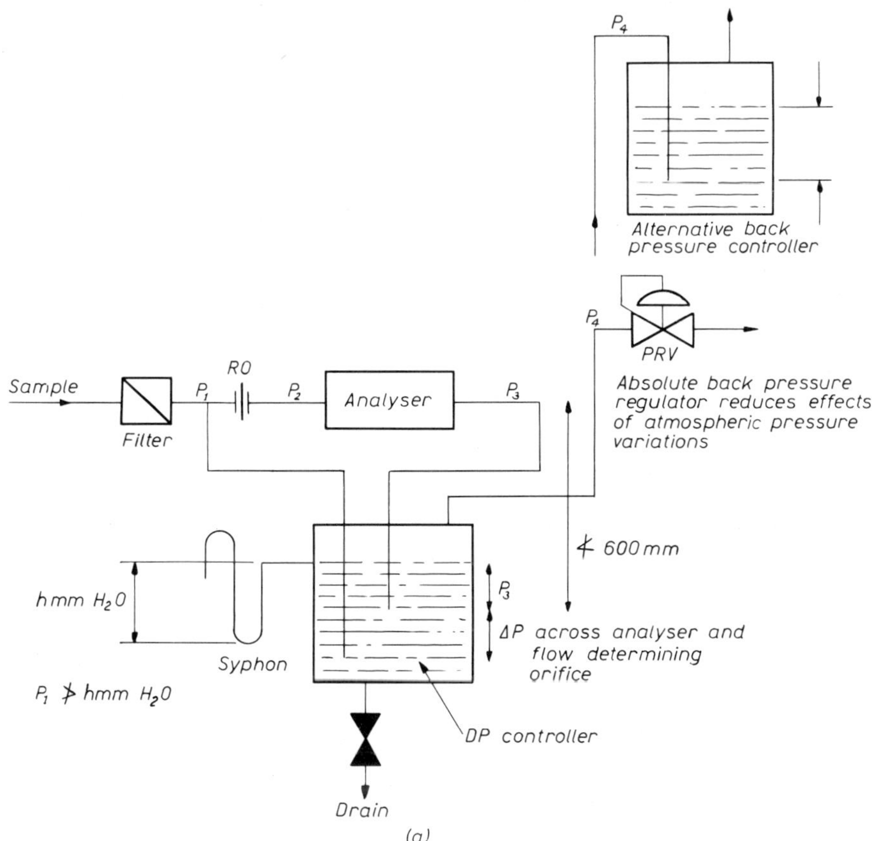

Fig. 8.62 (*a*) – Pressure and flow control. $P_3 - P_1 = 50$ mmH$_2$O, typically, at the required flow-rate.

(*b*) *Air aspiration.* In the air-aspiration method, an aspirator similar to that used in water aspiration is situated after the oxygen analyser (Fig. 8.63). A flow of air passes down the centre jet of the device and produces a low-pressure region in the venturi, which draws the sample gas from the sample probe through the analyser. It is simple to instal, and there is no contamination from water, but leakage of air into the system is possible.

(*c*) *Steam ejection at the cold end of the probe.* Here (Fig. 8.64) an open-ended sample tube protrudes into the flue. A steam ejector system is fitted at the cold end of the sample tube and a high-speed jet of steam is passed through it. This produces a low-pressure region in the centre of a venturi and causes the flow of flue gas out of the probe. Steam and sample-gas pass through the bottom of the jet into a cooler and separator unit; sample-gas then passes to the analyser. The use of desaturated steam does not introduce any dissolved oxygen into the sample stream, and the condensed steam dilutes sample-gas condensate, so reducing the effects of corrosion.

One disadvantage is that there is no scouring action inside the probe to prevent blockage. With suitably designed probes this system can be used at up to 2000°C.

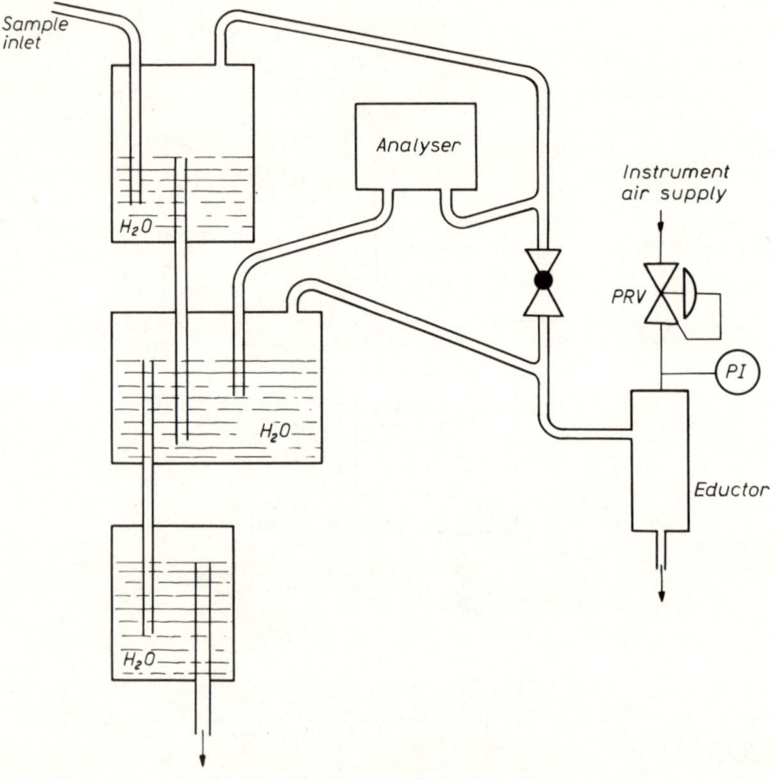

Fig. 8.62 (*b*) – Pressure drop control.

(*d*) *Steam ejection inside the flue* (Fig. 8.65). With this system a steam supply is taken through a tube right into the flue. This tube then bends through 180° and passes the steam through a jet into an open venturi. The resultant low pressure draws in flue gas which passes, together with the steam, through the neck of the venturi into an outlet tube. After the venturi, a pressure recovery occurs, and the mixture of steam and sample-gas passes down the sample lines to a sample separator. From the separator, the sample-gas passes through a centrifugal separator into the analyser. This system has many operating advantages.

   (i) The system is under positive pressure from inside the flue, so leaks in the sample line can only allow steam out, and not air in.

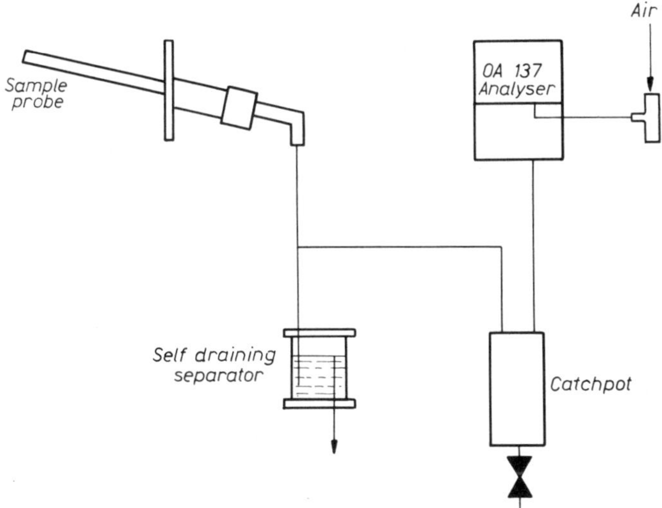

Fig. 8.63 – Air-aspiration method.

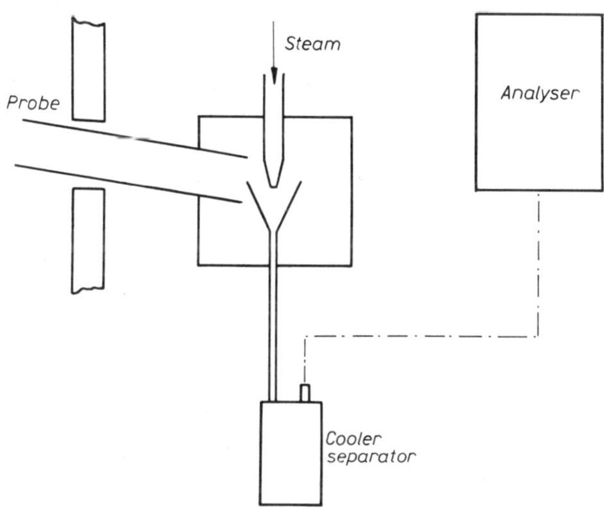

Fig. 8.64   Steam ejection outside the flue.

(ii) The high-speed steam jet scours the tube, preventing build-up of scale or dirt, etc.

(iii) The steam keeps the whole of the sample probe above the dew point and so prevents formation of corrosive condensate on the outside of the probe. This and the use of type 316 stainless steel or Hastelloy C make for long life and low maintenance cost.

(iv) The steam cools the probe and so allows it to be used at a much higher flue gas temperature than would be possible with a normal open-ended steel probe.

(v) When the steam/gas sample cools to the dew point the condensed steam dilutes the corrosive condensate present in the flue gas and reduces the risk of corrosion of the sample lines.

(vi) The system is intrinsically safe.

(vii) When used with the I.C.I.-designed steam probe (Fig. 8.65), it uses only 3.5 kg of steam per hour at between 10 and 20 bar ga.

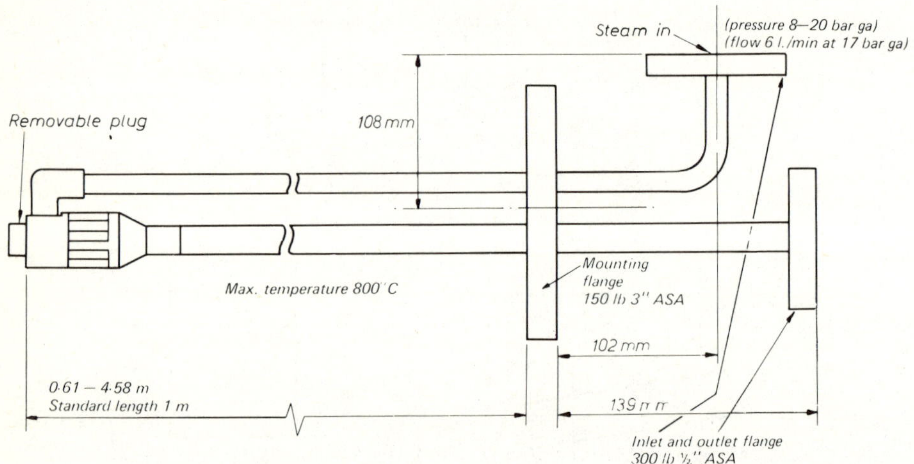

Fig. 8.65 – Servomex steam ejection probe, Type 195 (sample volume 0.15 l./m length).

(*e*) *The diaphragm-pump system* (Fig. 8.66). This is simple to instal and most of the disadvantages of the water-aspiration method are absent, but expensive explosion-proof motors may be required, and mechanical wear of the moving parts is inevitable. The working principle of the Siemens electromagnetic membrane pump is illustrated in Fig. 8.66 (*b*).

(*f*) *Rotary pumps.* Difficulties experienced with rotary pumps arise from leaks at the shaft glands which allow air into the pump and so produce errors. This problem has been overcome by using water-sealed glands. If these glands leak, water, not air, flows into the sample stream. High sample flow-rates are possible, but, as with diaphragm pumps, there are mechanically moving parts and wear is inevitable. Pumps are available with pneumatic motors and these are particularly useful in flammable atmospheres. Rotary pumps are usually preferred to positive displacement pumps because of the large impulses developed by the latter.

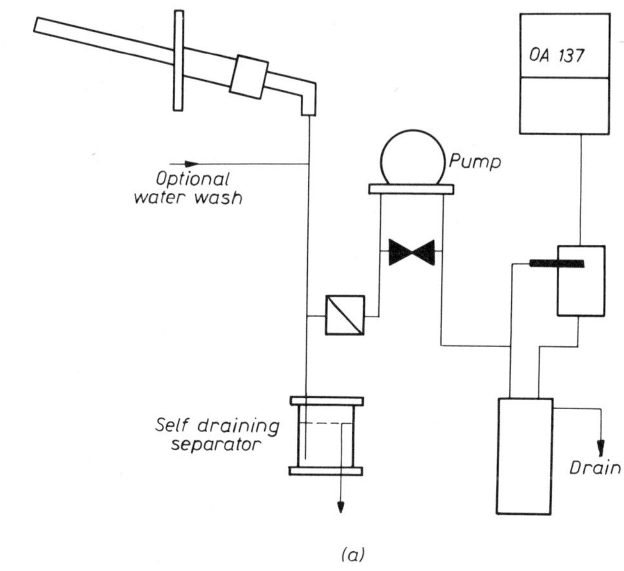

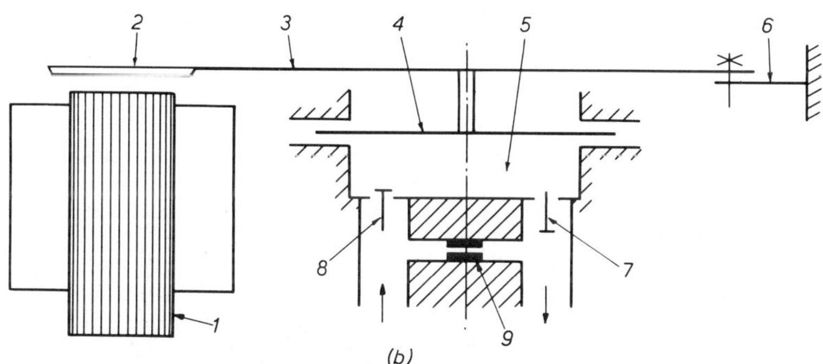

Fig. 8.66 (a) Diaphragm pump system. (b) Siemens electromagnetic membrane pump. 1, Electromagnet. 2, Soft-iron armature. 3, Vibrating arm. 4, Membrane (diaphragm). 5, Pump chamber. 6, Leaf spring. 7, Valve (pressure side). 8, Valve (suction side). 9, Shunting throttle to limit static suction.

8.9 COMMERCIAL GAS-SAMPLING PROBES

The Kent gas-sampling probes are illustrated in Figs. 8.67–8.70. The aerofoil probe (Fig. 8.67) is used where the gas to be analysed has a velocity of 6 m/sec or higher and contains considerable quantities of moist or dry suspended solids (e.g. rotating kilns used for cement manufacture).

The standard probe (Fig. 8.68) is made from mild steel (for up to 500°C) or stainless steel (for up to 800°C) and is fitted with a thimble-type primary filter, through which the gas passes before entering the sample system, ensuring a clean sample free from tars, solids, etc.

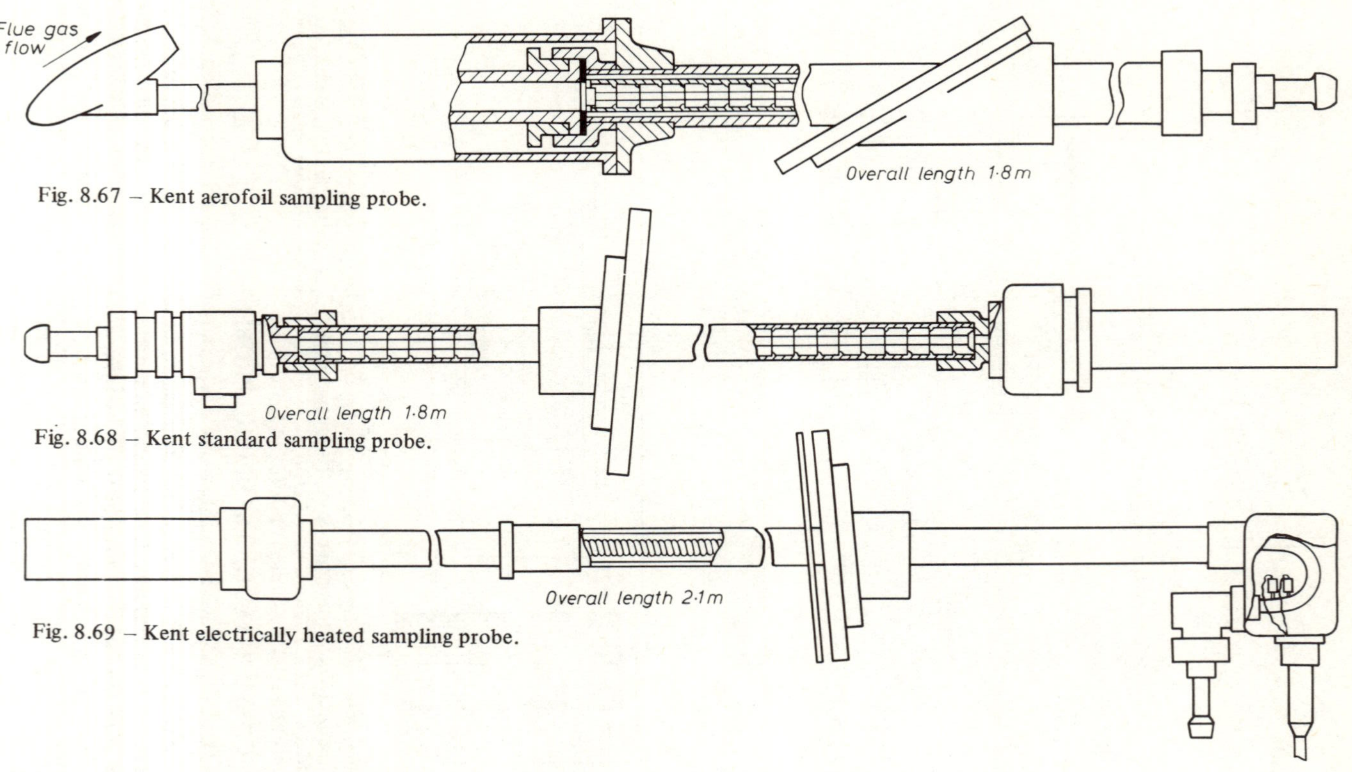

Fig. 8.67 – Kent aerofoil sampling probe.

Fig. 8.68 – Kent standard sampling probe.

Fig. 8.69 – Kent electrically heated sampling probe.

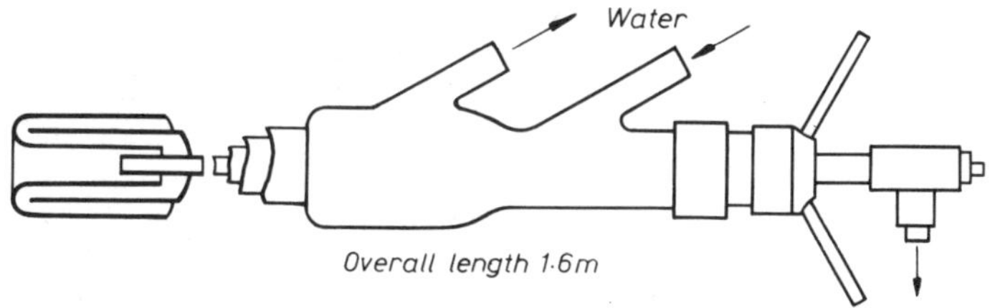

Fig. 8.70 – Kent water-cooled sampling probe.

The electrically-heated probe (Fig. 8.69) is used where the probe is mounted vertically and where there would otherwise be a danger of condensation occurring within the probe and causing corrosion (e.g Lancashire-boiler applications). In order to prevent blockage of the probe owing to failure of the heater supply, a direct-acting solenoid valve is included in the aspirator water-supply line. This valve is open only when the heating element in the probe is energized. The maximum working temperature is 500°C.

The water-cooled probe (Fig. 8.70) is used either to 'freeze' a sample to prevent further changes in gas composition after sampling or to permit sampling at temperatures in excess of those possible with uncooled metal tubes.

A Kent gas-sampling system is illustrated in Fig. 8.71.

In the Leeds and Northrup oxygen-analysing system (Fig. 8.72) the flue gas is drawn through the reverse jet probe by the steam ejector. The probe is continually flushed along its full length by water from a high-velocity jet which is provided by a nozzle mounted on the end of the probe.

The flue gas is cleaned by the jet condenser and centrifugal separator. In condensing, the steam scrubs out corrosive material and envelops all dirt particles to form droplets of water. The centrifugal separator removes the water–dirt–acid droplets and allows them to drain out through the float-operated level-control valve. The clean gas sample is discharged at the top of the separator at a positive pressure of up to 0.4 bar ga. Probe blockage is minimized by the high-velocity reverse jet. This self-flushing action keeps the probe clear.

The Leeds and Northrup 7892 assembly (Fig. 8.73) is designed to provide low-maintenance and fast-response gas sampling at high velocities under positive pressure from flues or ducts with gas temperatures up to 450°C. It is particularly suitable for dust laden gases. A steam-jet is used to pump the sample, the steam serving also to clean the sample gas.

The probe consists of a steam ejector mounted on the gas-inlet end of the sampling pipe which projects into the duct. Steam is supplied to the ejector through a pipe mounted parallel to and outside the gas-sample pipe. A tube inside the gas-sample pipe introduces water to condense the steam as it leaves the ejector.

The ejector creates a lower pressure at the nozzle outlet than the gas-sample pressure, thus aspirating the sample into the probe at this point. The steam then mixes with the sample. At the outlet of the ejector the mixture is under positive pressure. The condensing water is injected into the mixture at this point, condensing the steam and enveloping any dust particles present. The mixture then passes back to the probe outlet, where it passes to a separator (Fig. 8.73) which removes the dirt and water and provides a clean sample under positive pressure, moving at a velocity of approximately 15 m/sec.

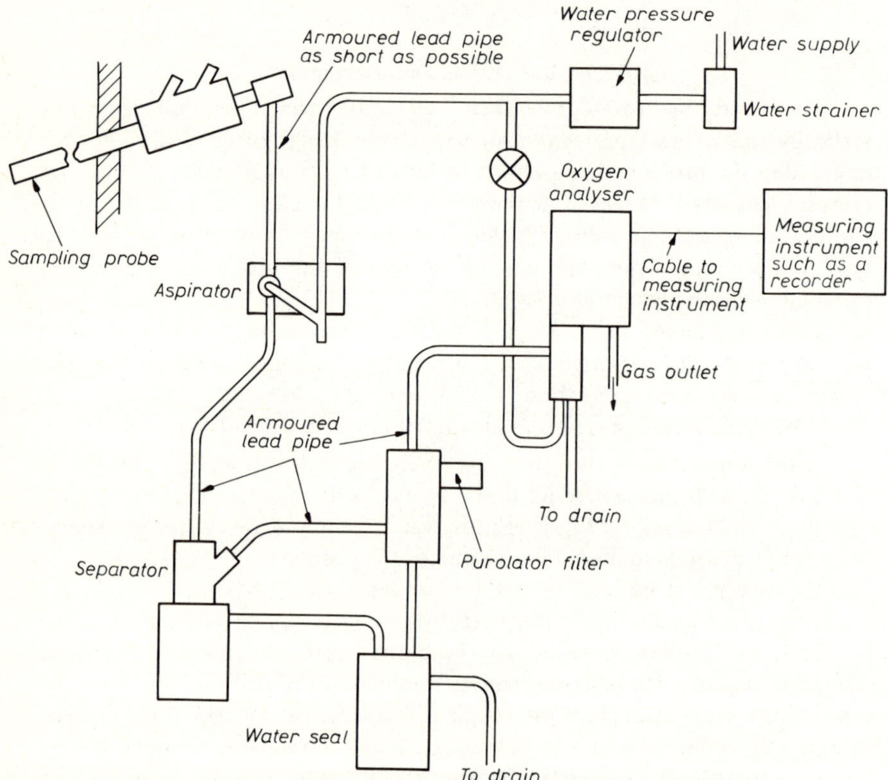

Fig. 8.71 – Typical Kent gas-sampling system using a water-cooled probe.

The Sprengel pump, which is illustrated in Fig. 8.74, is another method of aspirating the sample, washing it and supplying it to the analyser at a predetermined pressure.

A variation of this system which has been used with the Kent flue gas oxygen analyser is illustrated in Fig. 8.75. George Kent Ltd. is now a part of the British Brown Boveri Company.

In the Siemens probe system (Fig. 8.76), flue gas at 400°C with a 65°C dew point is cooled to 2°C so that the resultant dew point is always below the minimum ambient temperature, thus preventing condensation of water vapour in the analyser sample.

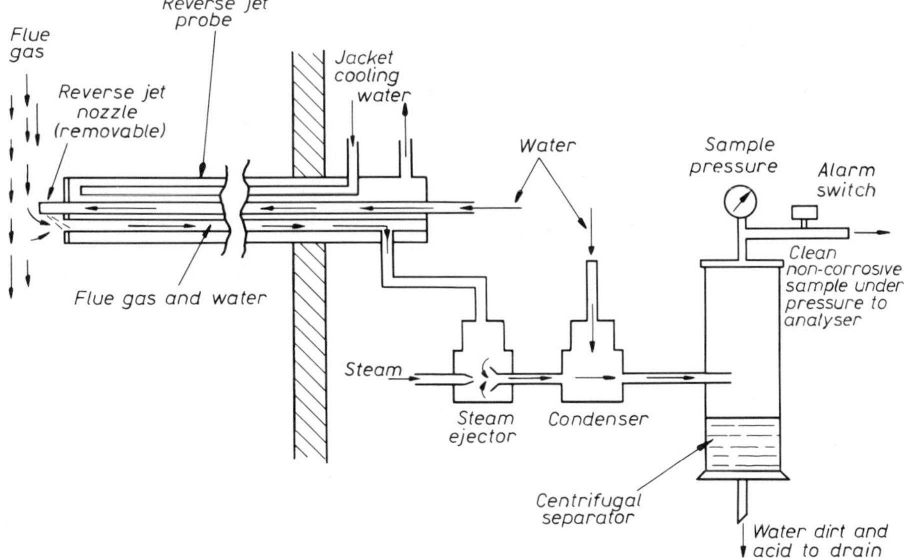

Fig. 8.72 – Leeds and Northrup oxygen-analyser sample system.

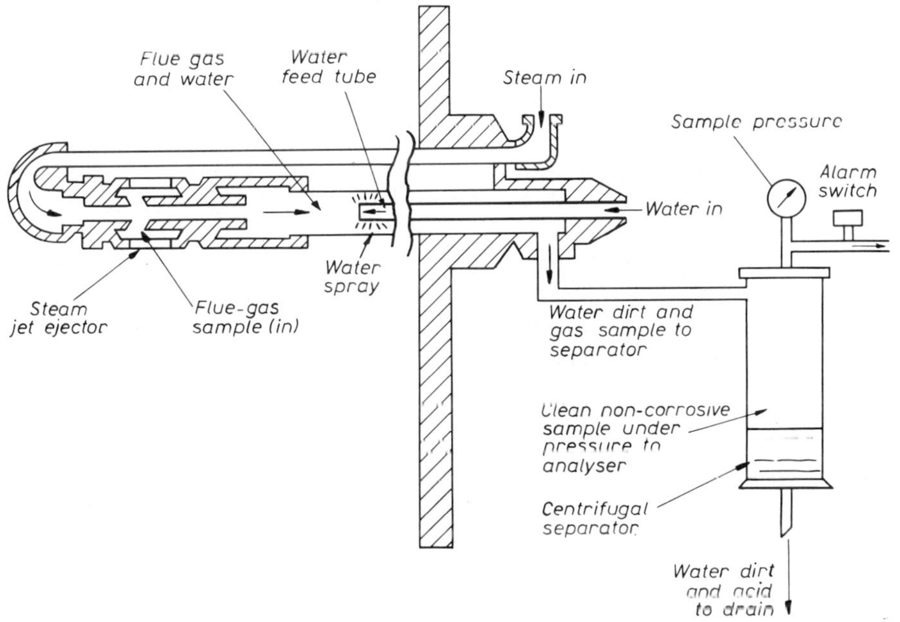

Fig. 8.73 – Leeds and Northrup 7892 assembly.

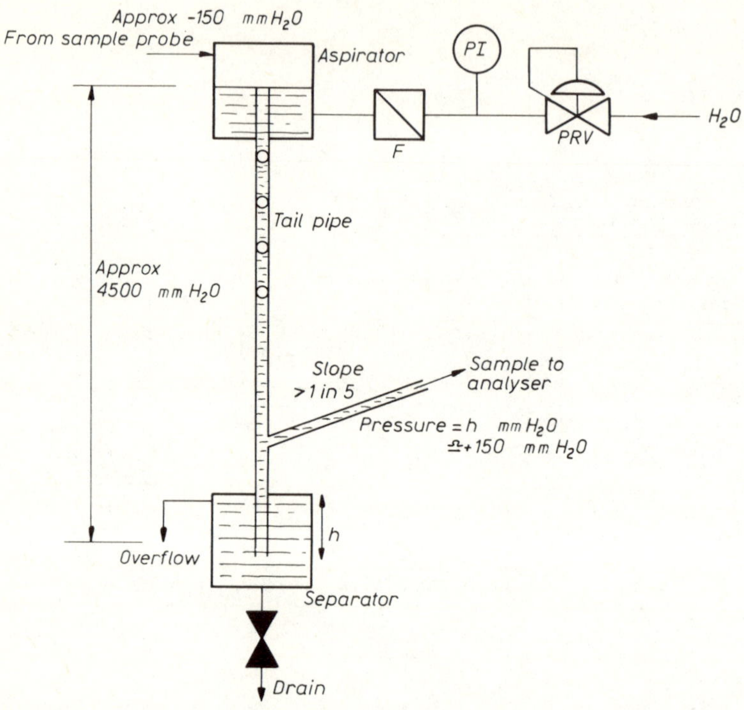

Fig. 8.74 – The Sprengel pump.

The temperature of the cooling water should be at least 5°C below the minimum ambient temperature. The system used with the Junkalor probe (Fig. 8.77) is similar to the Siemens system.

Two interesting Hartmann and Braun probes are illustrated in Figs. 8.78 (*a*) and (*b*). After being used for cooling, the water passes through a tangential nozzle. The water is thus discharged as a fine spray which washes the sample gas before it enters the probe. A water ejector is also built into the top of the probe.

The probe in Fig. 8.78 (*a*) is typically 1 m long and can be used at up to about 1400°C. A gas flow of 0.24–0.42 m³/hr may be obtained with water injected at 0.20–0.26 m³/hr and 5–8 bar ga.

Variation in amount of dissolved gas can be compensated by installing a temperature element. A totally enclosed coolant loop obviously has many advantages. Circulation may be forced or natural.

The probe in Fig. 8.78 (*b*) is typically 1.5 m long and can be used at up to 1500°C. A sample gas-flow of 2 m³/hr may be obtained with injection of 0.5–3 m³/hr of water at 2–6 bar ga. The rupture disk is usually set to burst at about 10 bar ga.

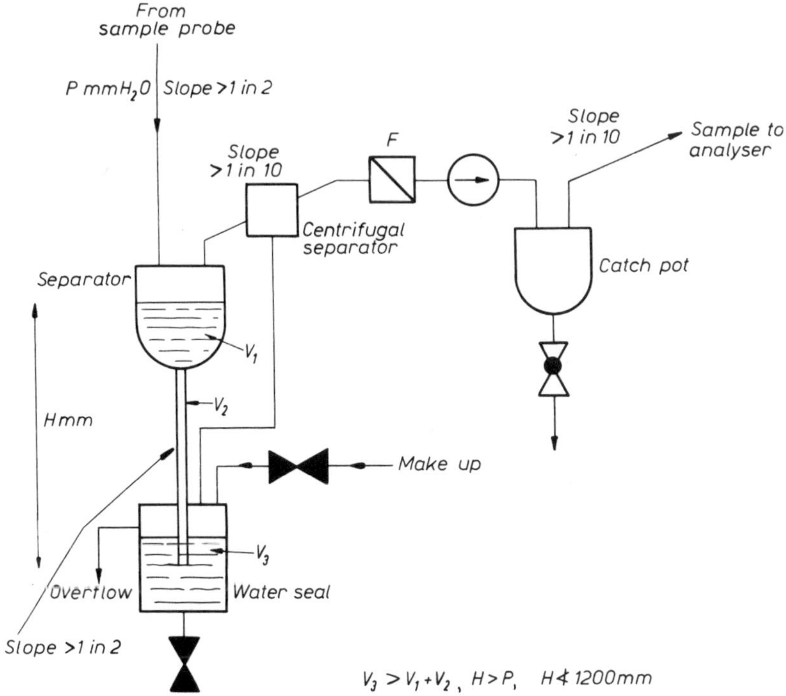

Fig. 8.75 — The Sprengel pump as used with the Kent flue gas oxygen analyser.

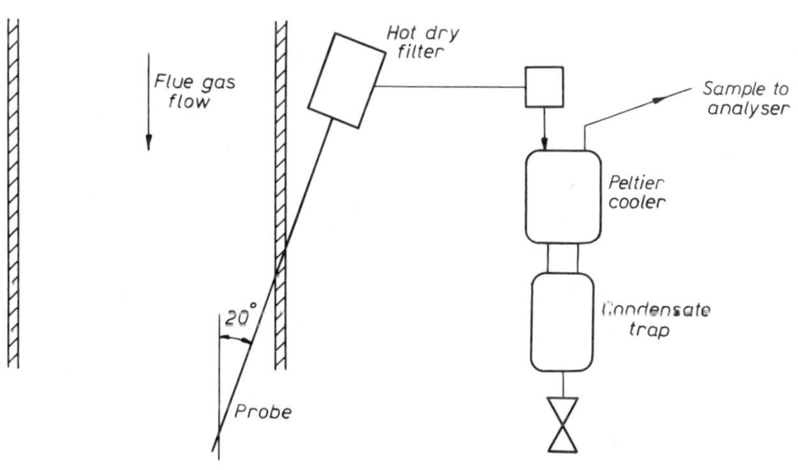

Fig. 8.76 — Siemens probe and sample cooling system.

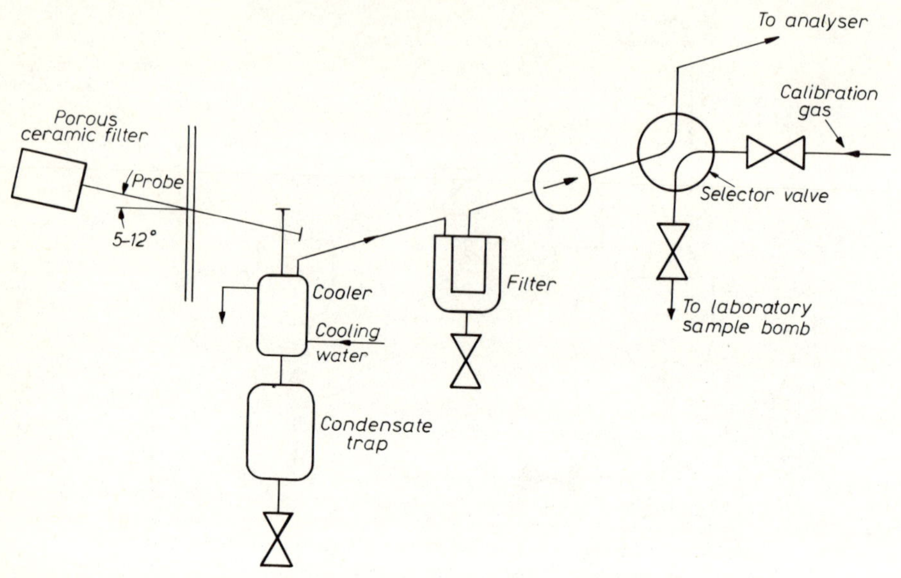

Fig. 8.77 – Junkalor probe system.

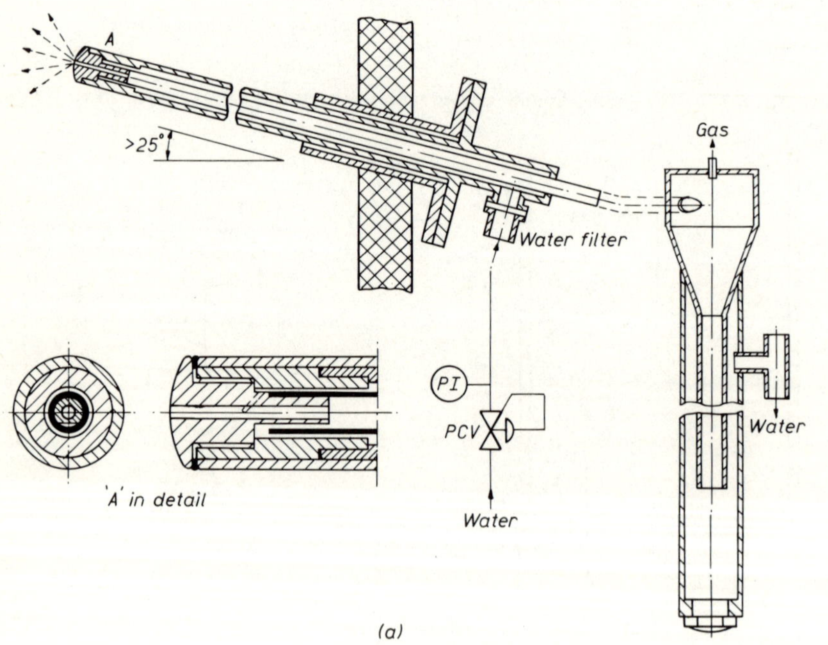

(a)

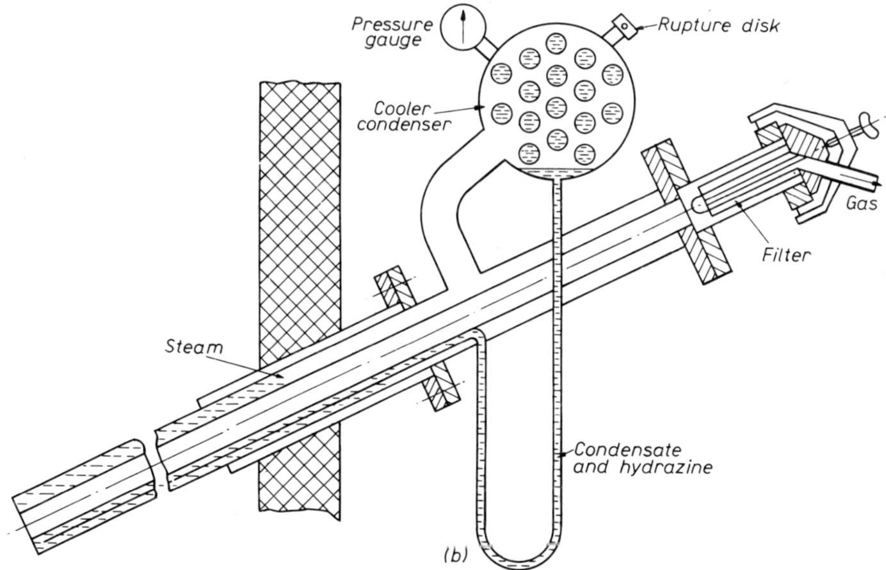

Fig. 8.78 (*a, b*) – Two types of Hartmann and Braun probe.

Cleaning by back-flushing with compressed air or nitrogen is recommended. Unburnt gases, if not sufficiently cooled, can burn at a temperature dependent on the catalytic action of the material.

Silicon carbide may be used at up to 450°C and quartz and aluminium oxide at up to 650°C.

The Servomex* steam ejector system is shown in Fig. 8.79. (The Servomex steam ejector probe, illustrated in Fig. 8.65, was described earlier along with its advantages).

8.10 MULTIPLE SAMPLE SYSTEMS

In many cases it is desirable to monitor a number of sample points without requiring a continuous record to be made of each point. It may not be economic to have a separate oxygen analyser for each point and, if this is the case, sequential automatic sampling can be used.

In a typical six-way automatic-sampling system using steam-ejector probes for the flue-gas sampling and electromagnetic solenoid valves to select the sample stream, the six solenoid valves are controlled by a six-point strip-chart recorder. The wiring of the recorder is changed so that it has a single input connection, but the switch normally used to select different inputs is now used to select each of the solenoid valves in sequence. Owing to the speed of response of the analyser in the system, stream-switching can take place every 30 sec. In this way, each sample point is selected every 3 min and a coloured dot stamped on the recorder chart. These dots are sufficiently close together to produce

*Now Taylor-Servomex.

the effect of a continuous line, although obviously with this system, high-speed transients would not be detected. Using the recorder to control the solenoid valves ensures that there is no risk of sample streams and record being out of step.

An alternative method is to use an external timer to select each of the solenoid valves in sequence, the output from the analyser being presented to a single pen recorder. With this system, it is normal to sample from each point for between 5 and 10 min. Any time interval less than this makes the record unreadable.

Automatic sequential sampling is extremely useful in determining the profile of an oxygen mixture in a large kiln, where the conditions do not normally change rapidly, and yet where a large number of sample points is necessary in order to determine the oxygen curve.

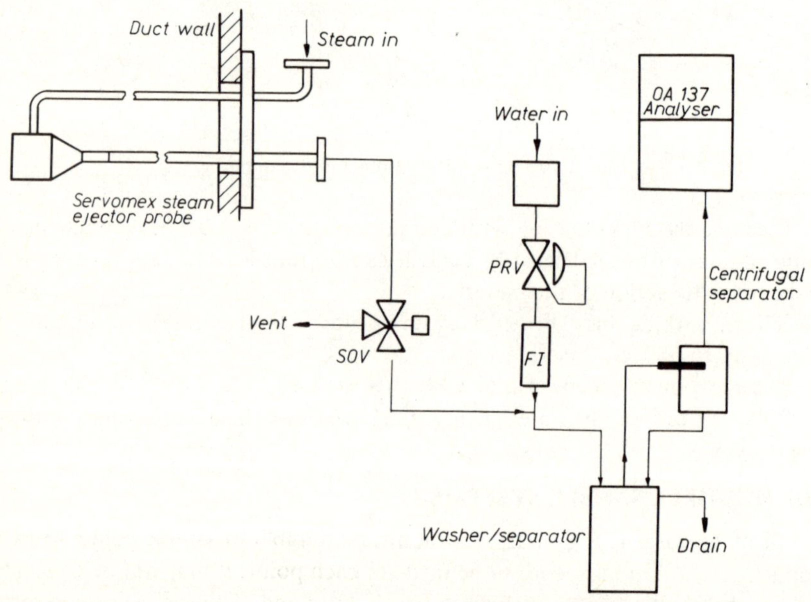

Fig. 8.79 – Servomex steam-ejector system.

An interesting method has been developed by the CEGB, at Bristol (Fig. 8.80). This method uses a large number of sample points across the flue in a power station. Each of these points is connected to a solenoid valve. Each solenoid valve in turn is energized to allow the sample to flow through the analyser.

The interesting feature of this system is that normally the maximum sample flow through the Servomex oxygen-analyser measuring cells is 150 ml/min. However, with this system, a sample flow of 4 l/min is allowed through the measuring cell for a time of 4 sec. At the end of this time, the sample flow is stopped; 1 sec is allowed for the system to reach equilibrium and then a measurement is made. With this system it is therefore possible to sample with only 5-second intervals between samples.

Where it is necessary to monitor the oxygen content in a large flue, such as is found on a power station, samples are extracted from different points in the flue and are mixed together before feeding into the oxygen analyser. In this way, the analyser sees the average composition of gas over the whole area of the flue and avoids possible errors due to layering.

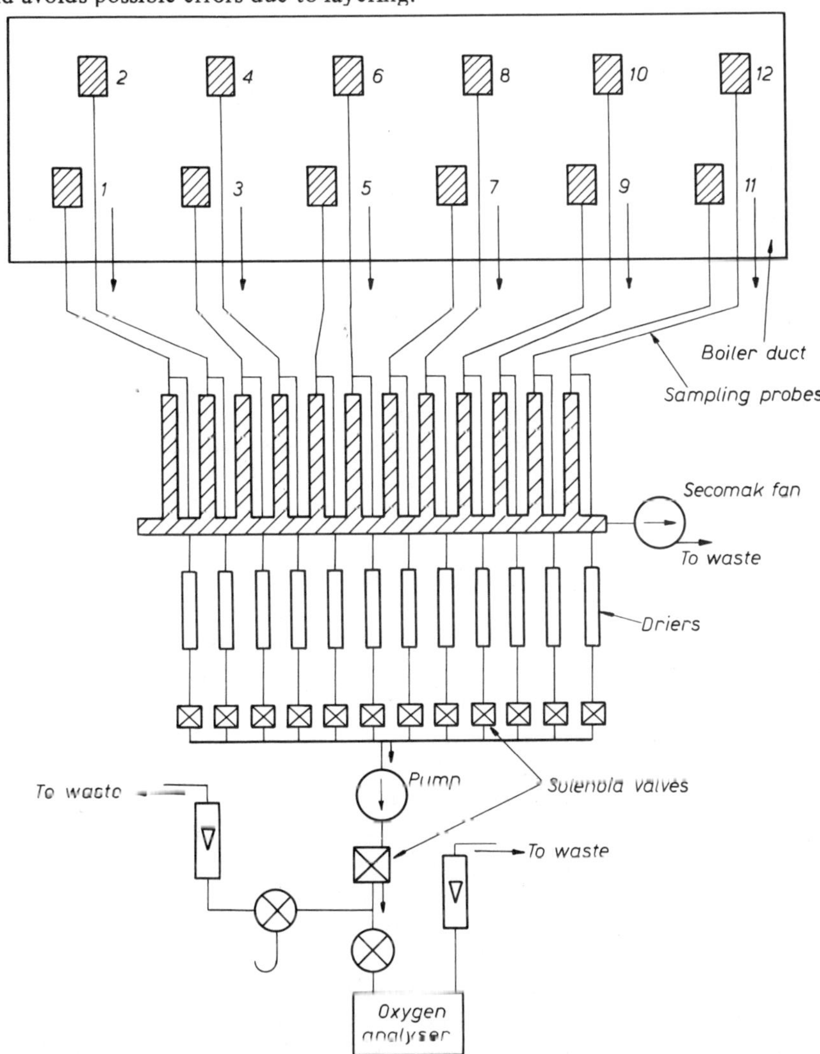

Fig. 8.80 – Automatic sampler developed by CEGB, Bristol.

The most common method of sample averaging is to use a number of complete sampling systems, and then to feed the dry, clean gas through individual flowmeters before mixing it and passing it to the analyser (Fig. 8.81).

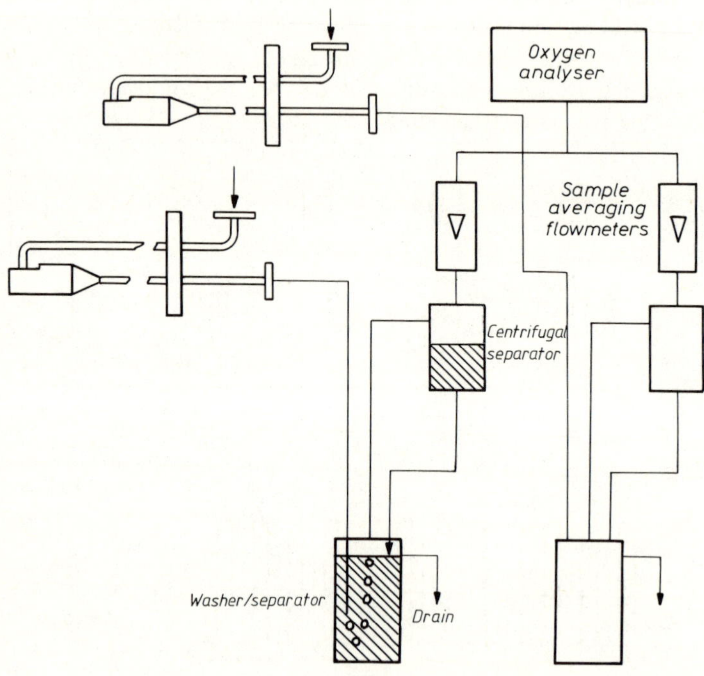

Fig. 8.81 – Oxygen analysis of the average of two samples.

With normal flowmeters it is impossible to measure the sample flow immediately after each sample probe, as the sample gas contains moisture and dust.

An interesting system has been devised by the Fuel Efficiency Dept., CEGB, Croydon, to overcome the problem of sample averaging with a portable unit. Individual sample lines are brought from each sample point down to a mobile analysis unit. Each sample tube is connected to a hand valve and then to a common wash bottle. After the wash bottle there is a single diaphragm pump which pushes the mixed sample into the oxygen analyser (Fig. 8.82). In order to set the individual flow of each sample, a thermistor is inserted in each of the sample lines before the mixing unit, the tip of the thermistor being in the centre of the sample tube. With this system, the flow of gas can be measured directly, as a change in the temperature of the thermistor, with an error of ± 5%. It is interesting here to note that the condensate running down the inside surface of the tube does not appear to have any effect on the reading of the thermistor.

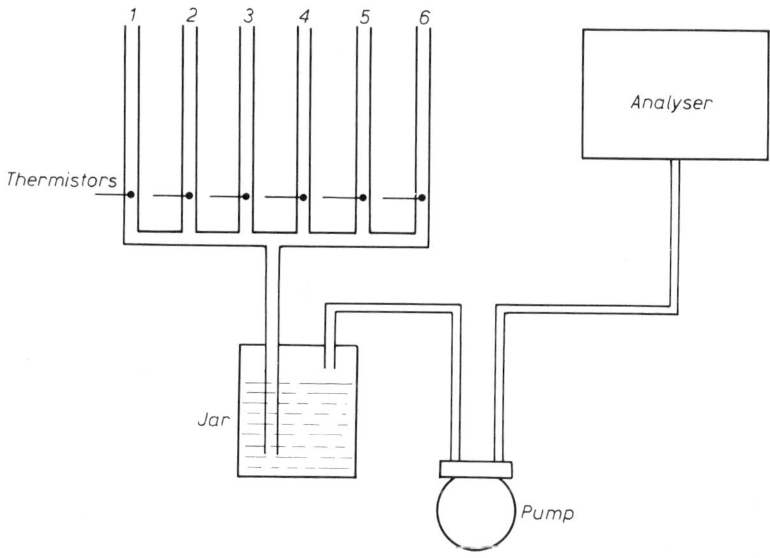

Fig. 8.82 – Portable sample averaging unit developed by CEGB, Croydon.

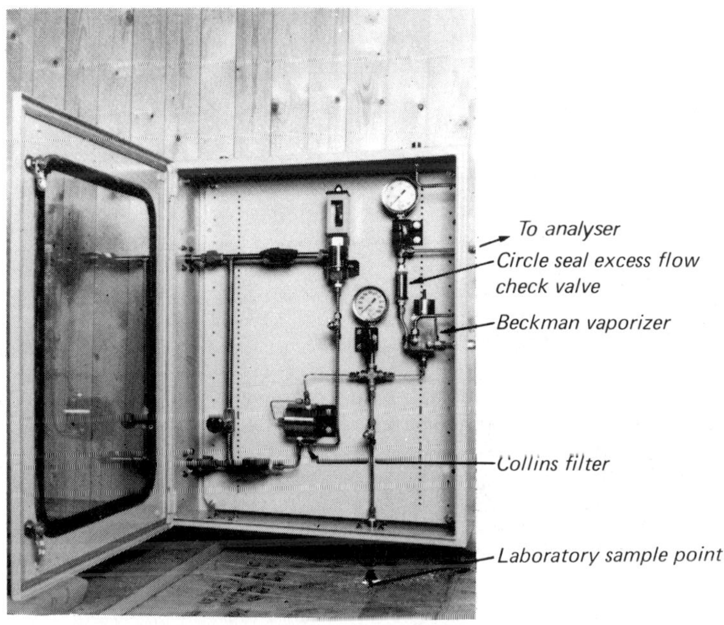

Fig. 8.83 – External fast loop sample conditioning system vaporizing a liquid sample. Designed by Fluor and made by Control et Applications – Sterwech.

A more complex method of sample averaging is to use an individual oxygen analyser for each sample point and then to feed the outputs from each of these analysers into a computer. This system has the advantage that if one point in the flue shows an exceptionally high oxygen content, this can be detected by the computer as a possible fault condition. This fault condition could arise from a large leak in the wall of the flue.

Two external fast-loop sample-conditioning systems shown in Figs. 8.83 and 8.84 are well designed and made, but need to be revised when new systems are made.

In the single-stream system, the fast-loop liquid flow should be downwards.

In the three-stream system, the sight-flow glasses could be relocated to the system inlets (Fig. 8.85) so that they may also be used to indicate sample-line flow before the system itself is connected and to allow maintenance staff to observe whether the sample is clean or not.

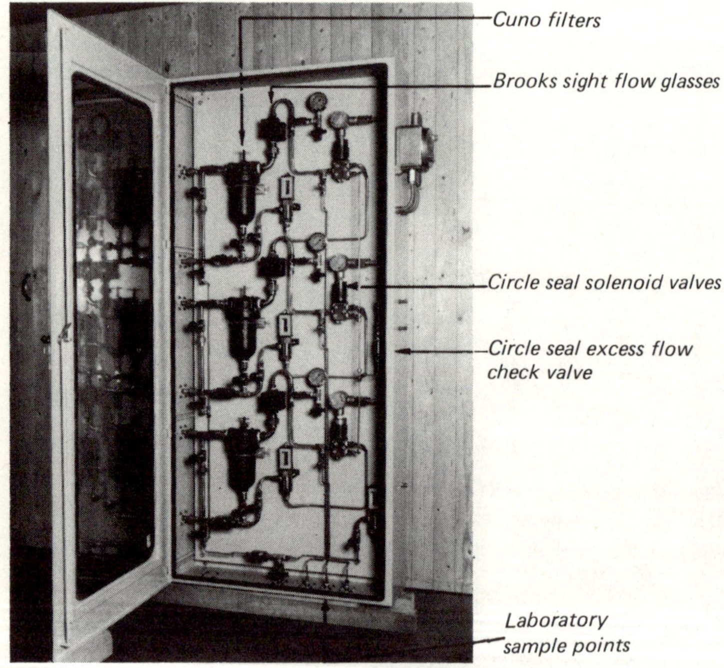

Fig. 8.84 — External fast loop sample conditioning system sequentially selecting one of three liquid samples. Designed by Fluor and made by Control et Applications — Sterwech.

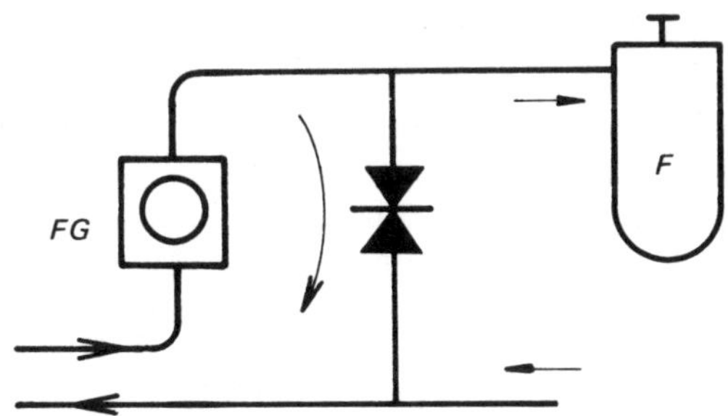

Fig. 8.85 — New location of the sight-flow glass.

REFERENCES

[1]   L. F. Moody, *Trans. Am. Soc. Mech. Eng.*, 1944, **66**, 671.
[2]   *Flow of Fluids Through Valves, Fittings and Pipe*, Crane, London, Imperial Ed., 1977, Metric Ed., 1979.
[3]   E. A. Houser, *Principles of Sample Handling and Sample Systems Design*, Beckman Instruments.
[4]   W. E. Lobo, L. Friend and G. T. Skaderpas, *Ind. Eng. Chem.*, 1942, **34**, 821.
[5]   *Engineering Data Book*, 9th Ed., Gas Processors Suppliers Association, Tulsa, Oklahoma, 1972, Section 18.

CHAPTER 9

Analyser enclosures - the fire, toxic and radioactive hazards

9.1 ANALYSER HOUSE SIZE AND CONSTRUCTION

Analysers are usually weather-proof and explosion-proof or flame-proof. They can thus be mounted in the open without any further protection, but this is not normal practice, especially in corrosive atmospheres. If it is necessary to service analysers in bad weather, it is better to mount them in an enclosure such as a walk-in cubicle (for individual analysers) or an analyser house (for several analysers).

The internal dimensions of an analyser house depend very much on the number and type of analysers in the house. Typical dimensions are approximately 3 m wide × 3-6 m long × approximately 2.5 m minimum height above floor level. The height is sometimes increased by a further metre so that piping and wiring can be run above the analysers.

The clearance between rows of analysers should be approximately 1.5 m. This clearance is more than adequate for maintenance and for safe exit during a fire. For the latter reason this central space should be kept clear of any obstacles. The clearance between analysers along the side of the wall of the analyser house should be adequate for all interconnecting piping, conduit, valves, terminal boxes etc. and for maintenance access. If the analyser is handling a toxic fluid, it is advisable to provide additional clearance around it with good local ventilation and possibly one or two gas monitors to detect leakages.

Many analyser houses are built of bricks or of concrete blocks. This type of construction tends to reduce the temperature variations within the house. Such houses are thus common in countries (such as those in the middle-east) where it is very hot during summer daytime. Prefabricated houses are usually cheaper than brick or block houses. These are made of concrete, reinforced glass fibre and insulated sheet-steel or aluminium. They may be piped and wired and then shipped to site as a complete tested package which requires very little expensive field installation.

The cost of the cubicle or analyser house, piping, blowers, etc., is normally about 20-100% of analyser cost, depending on the number of analysers in the enclosure.

The concrete floor of analyser houses absorbs liquids and thus creates a resultant hazard. The concrete floor should therefore be coated with a fire-resistant, sample-liquid-resistant, non-slippery material, which is hard and will not crack easily. Vitreous clay ceramic tiles are most suitable. These should be fixed to the concrete floor with a non-porous cement. The floor should slope gently towards a drain so that puddles do not gather. The floor beneath a liquid analyser should be fitted with a curb to prevent the spread of spillage.

Analyser houses beyond a certain minimum size ought to have at least two doors so that anyone caught there during a fire will have a reasonable chance of getting out without serious burns. The doors should be located at the two ends of the house and should open outwards to a clear area so that personnel can escape safely from the house. Both doors should be fitted with armoured glass windows or lenses to enable almost complete internal inspection from outside. They should be capable of being locked only from the outside and should be fitted with emergency push-bars on the inside. A typical design is shown in Fig. 9.1.

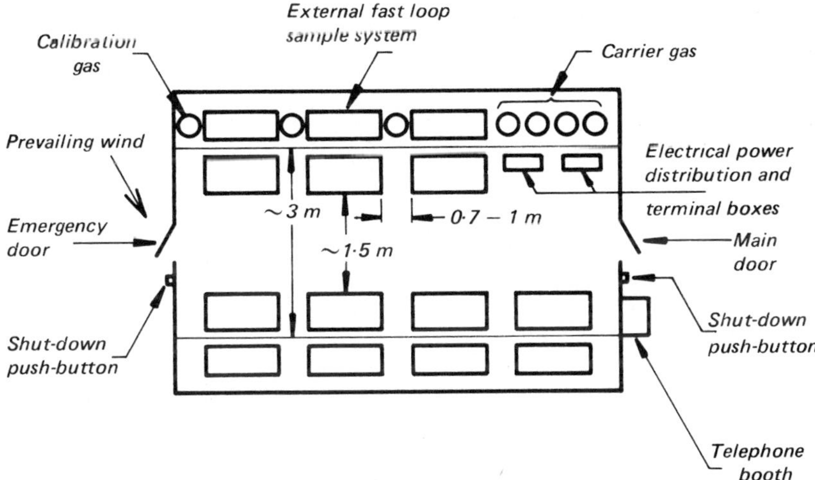

Fig. 9.1 – A typical analyser house.

Many analysers require a water supply, and the water is normally required to be clean and cold. If the only suitable supply is potable water, a break tank may have to be installed to prevent any possibility of back-diffusion of the process fluid into the potable water supply. The break tank is often an enclosed atmospheric tank mounted on top of the analyser house: the water inlet is regulated by a float valve.

Analysers in analyser houses should be vented (safely) to remove toxic or flammable sample or reagent, etc., and drained to a line with a sufficiently small back-pressure outside the analyser house (Fig. 9.2). Analysers which must be vented separately may be vented by a separate ½ in. pipe supported by the common 2-in. vent pipe. The vent discharge should be at least 3 m above the roof of the analyser house. Areas within 1.5 m of the vent discharge are designated as Zone 1. The resultant Zone 2 area extends to about 7.5 m (see Section 9.3).

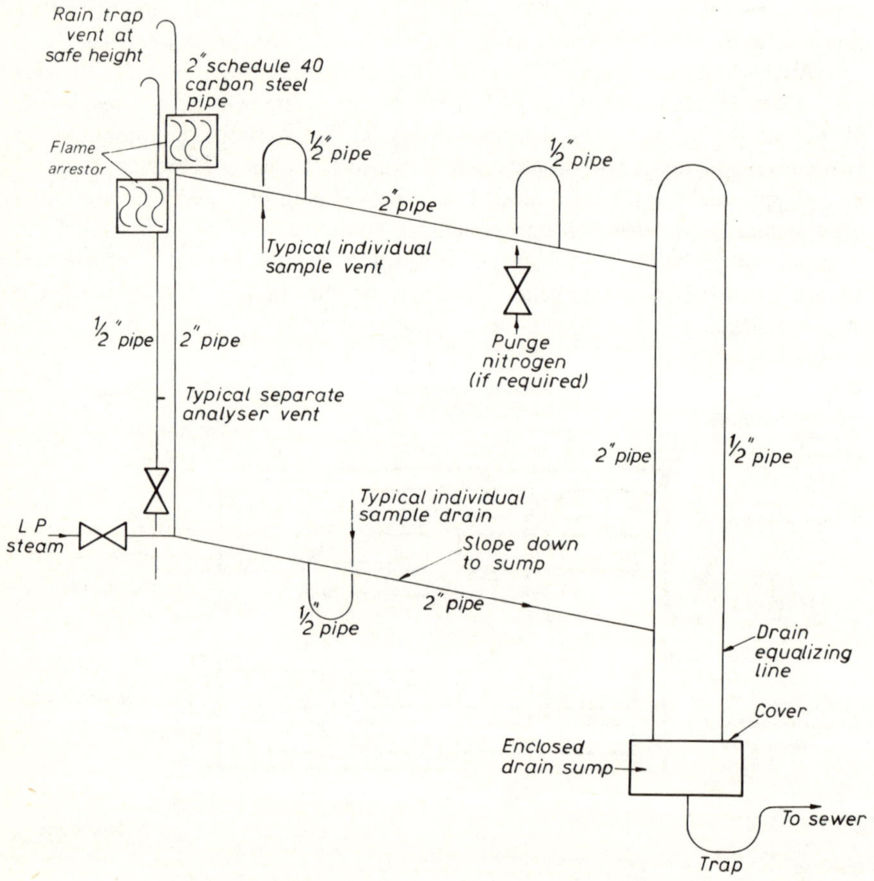

Fig. 9.2 – Analyser-house vent and drain system.

A low pressure steam connection should be made to the drain down to the sump, to ensure that heavy components flow to drain and that most flammable fluids are vented, and flame arrestors should be fitted in the drain system.

A flame arrestor may consist of a bundle of tubes in which the heat released by combustion is absorbed by the tube walls. The flame velocity is decreased as the diameter of the tubes is decreased to about 1–2 mm, when they completely prevent the passage of the flame, irrespective of velocity. Flame arrestors may also be made from alternate flat and corrugated sheets.

A wire screen of sufficiently fine mesh will not permit a flame to pass through it, but it must remove a large quantity of heat by conduction to its periphery in order to prevent the screen itself from attaining a high enough temperature to ignite the gas.

Flame arrestors should always be used for low flash-point, highly volatile fluids with flash points less than about 100°C. Open vents require flame arrestors, but closed vents and drains do not require flame arrestors.

9.2 VENTILATION AND AIR CONDITIONING

In a normal analyser house about 3–4 air changes per hour may be produced by natural ventilation; this may be increased slightly by making use of the wind and natural suction at the top of a sloping roof.

If more air changes are required, a forced draught fan or blower should be used. Induced draught fans should be used only with very great caution.

Heat is lost by conduction through the walls, roof and floor and in the air exhausted. This latter loss is often not very significant when only natural ventilation is used.

The air should be heated by heaters mounted close to the inlet ports; it will also be heated by heat dissipated by the analyser and other equipment within the enclosure and by solar irradiation.

The heat transfer through a wall is given by

$$Q_w = \Delta T_w . K_w . A_w \quad \text{kcal/hr}$$

$$= \Delta T_w . U_w . A_w \quad \text{W}$$

where ΔT_w = temperature difference across the wall (°C), K_w = heat transfer coefficient in kcal.m^{-2}. hr^{-1}. °C^{-1}, A_w = surface area of the wall (m^2), and U_w = thermal transmittance of the wall (W.m^{-2}, °C^{-1}).

For example, for an insulated steel wall, the heat transfer coefficient is given by

$$\frac{1}{K_w} = \frac{1}{20} + \frac{1}{\alpha_0} + \frac{t_i}{\lambda_i} + \frac{t_s}{\lambda_s} + \frac{1}{\alpha_1}$$

where α_0 = outside surface convection factor $\cong$ 20 kcal.m^{-2}.hr^{-1}. °C^{-1} for air with a velocity of the order of 4 m/sec, α_1 = inside surface convection factor $\cong$ 7 kcal.m^{-2}.hr^{-1}. °C^{-1} for air, t_s = thickness of steel wall, typically 0·002 m, t_i = thickness of insulation $\cong$ 0·04 m, λ_s = thermal conductivity of steel = 62 kcal.m^{-1}.hr^{-1}. °C^{-1} and λ_i = thermal conductivity of insulation = 0·04 kcal. m^{-1}.hr^{-1}. °C^{-1}. (For aluminium λ_a = 180 kcal.m^{-1}.hr^{-1}. °C^{-1}).

$$\therefore \quad \frac{1}{K_w} = \frac{1}{20} + \frac{0·04}{0·04} + \frac{0·002}{62} + \frac{1}{7} = 1·19$$

$$\therefore \quad K_w = 0·84 \text{ kcal.m}^{-2}.\text{hr}^{-1}. °C^{-1} .$$

(Note 1 kcal . m^{-2} . hr^{-1} . °C^{-1} $\equiv$ 1.16 W . m^{-2} . °C^{-1}).

If insulation is not used, $K_w = 5$; thus it is advisable to use insulation. For a brick wall of thickness 0·19 m and thermal conductivity 0·75 kcal.m^{-2}.hr^{-1}.°C^{-1}.

$$\frac{1}{K_{BW}} = \frac{1}{20} + \frac{0·19}{0·75} + \frac{1}{7} = 0·45$$

$$K_{BW} = 2·2 \text{ kcal.m}^{-2}.\text{hr}^{-1}.°\text{C}^{-1}$$

For glass windows (6 mm thick)

$$\frac{1}{K_{GW}} = \frac{1}{20} + \frac{0·006}{0·7} + \frac{1}{7} \cong 0·20$$

$$K_{GW} \cong 5·0 \text{ kcal.m}^{-2}.\text{hr}^{-1}.°\text{C}^{-1}$$

For a concrete floor with no cavity beneath the floor,

$$\frac{1}{K_{cf}} + \frac{1}{\alpha_G} + \frac{t_c}{\lambda_c} = \frac{1}{5} + \frac{0·20}{1} = 0·4$$

$$K_{cf} = 2·5 \text{ kcal.m}^{-2}.\text{hr}^{-1}.°\text{C}^{-1}$$

where subscript c refers to concrete and subscript G to ground.

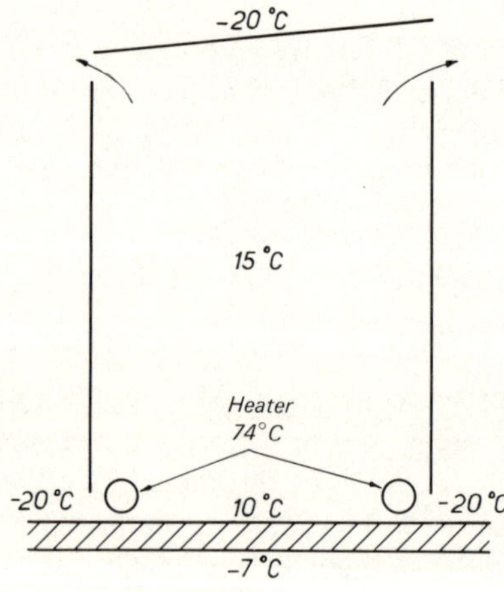

Fig. 9.3 – Typical analyser-house temperature distribution during a hard winter.

As an example to illustrate the various calculations involved, consider an analyser house 1·2 m wide × 1·2 m long × 2·4 m high with insulated steel walls, an uninsulated steel door, an insulated roof (K_R = 1·45 kcal.m^2.hr^{-1}.°C^{-1}) and a concrete floor.

Typical values of temperature during a hard winter are shown in Fig. 9.3 and the temperature differences will be:

$$\Delta T_{\text{roof}} = 15 - (-20) = 35°C$$

$$\Delta T_{\text{walls}} = 15 - (-20) = 35°C$$

$$\Delta T_{\text{floor}} = 10 - (-7) = 17°C$$

$$\Delta T_{\text{heater}} = 74 - (-20) = 94°C$$

Heat loss by air convection through the roof slots is given by

$$Q_{\text{conv}} = C_p \times N \times V \times T_R$$

where C_p is the specific heat of air $= 0\cdot31$ kcal.m^{-3}.°C^{-1}, N is the number of air changes per hour, V is the volume of the analyser house (m^3), T_R is the temperature at the top of the house (°C), so $Q_{\text{conv}} = 0\cdot31 \times 4 \times 3\cdot46 \times 15 \cong 64\cdot4$ kcal/hr.

The heat lost by conduction is given by

$$Q_{\text{cond}} = Q_{\text{walls}} + Q_{\text{roof}} + Q_{\text{door}} + Q_{\text{floor}}$$

$$= \Delta T_w \cdot K_w \cdot A_w + \Delta T_r \cdot K_r \cdot A_r + \Delta T_d \cdot K_d \cdot A_d + \Delta T_f \cdot K_f \cdot A_f$$

$$= 35 \times 0\cdot84 \times 3 \times 2\cdot88 + 35 \times 1\cdot45 \times 1\cdot44 + 35 \times 5 \times 2\cdot88 +$$

$$17 \times 2\cdot5 \times 1\cdot44$$

$$\cong 254 + 73 + 505 + 61$$

$$= 893 \text{ kcal/hr}$$

To this 10% is added to allow for extreme conditions of temperature, rain or wind, and plant deterioration.

$$\therefore \qquad Q_{\text{cond}} = 893 + 89\cdot3 \cong 982 \text{ kcal/hr}$$

$$\text{Total heat loss} = Q_{\text{conv}} + Q_{\text{cond}}$$

$$= 64.5 + 982 \cong 1047 \text{ kcal/hr}$$

$$= \text{total heat input required}$$

The electrical energy dissipated by an analyser and its associated components is typically 750 W $= 645$ kcal/hr $= Q_E$. Therefore, neglecting the effect of solar radiation, the required heat input from the steam heater is

$$Q_s = Q_{\text{conv}} + Q_{\text{cond}} - Q_E$$

$$= 1047 - 645 = 402 \text{ kcal/hr}$$

$$= C_p \times N \times V \times (T_H - T_I)$$

where $T_H =$ heater temperature and $T_I =$ temperature of inlet air.

$$402 = 0\cdot31 \times 4 \times 3\cdot46 \times (T_H - T_I)$$

$$T_H = \frac{402}{4\cdot26} - 20 = 94 - 20 = 74 \text{ °C}$$

This is the temperature of the air leaving the heater and is rather higher than is normally desirable, illustrating one of the limitations of natural ventilation.

Using steam at 7 bar pressure, with a latent heat (H_s) of 500 kcal/kg, the steam flow requirement is given by

$$W_s = \frac{Q_s}{H_s} = \frac{402}{500} \cong 0\cdot8 \text{ kg/hr}$$

In fact it will be slightly greater owing to the temperature drop across the heater wall.

If a finned ½-in. (12-mm) nominal bore pipe is used, the pressure drop will be negligible. Finned copper pipe has much better heat transfer properties than steel and it may be used up to about 150°C provided that the surface temperature is not greater than 80% of the autoignition temperature (°C) of any flammable gas handled by the analyser.

A temperature-control valve should be used when there are rapid and wide ambient temperature variations and wide variations in dissipation from electrical equipment. Cooling during hot summers is often desirable but seldom done, other than by ventilation.

Explosion-proof air conditioners are available, but they can occupy a large amount of space in the analyser house. In order to minimize the size of the air conditioner, the fresh air intake must be minimized. An air-change rate (make-up) of 5–10 per hr is typical.

Sufficient ventilation can produce adequate apparent cooling of personnel, owing to evaporative cooling of the skin.

In many countries producing crude oil, the maximum summer temperature in the shade can be about 50°C, which produces too high a temperature in the analyser house for most analysers and for effective maintenance by most people. Air conditioning is thus necessary, but it is not always used, because some operating companies prefer the cost of the increased analyser maintenance and can accept the decreased availability.

It is not difficult to show that an analyser house cannot be air-conditioned economically if the purge-rate is more than about 10 changes/hr. Thus, in hot climates most plant analyser houses must be Zone 2 areas (see next section).

Especially in sandy areas, it is best to pressurize the analyser house with air from a clean and, if possible, safe location by the use of a roof-mounted blower (Fig. 9.4). The air should then be distributed throughout the analyser house as evenly as possible. Finally, the air should be vented from the house through louvres near the roof and the floor. Check vents should be used where possible for it is otherwise very difficult to ensure that a high wind does not blow dirty air into the analyser house.

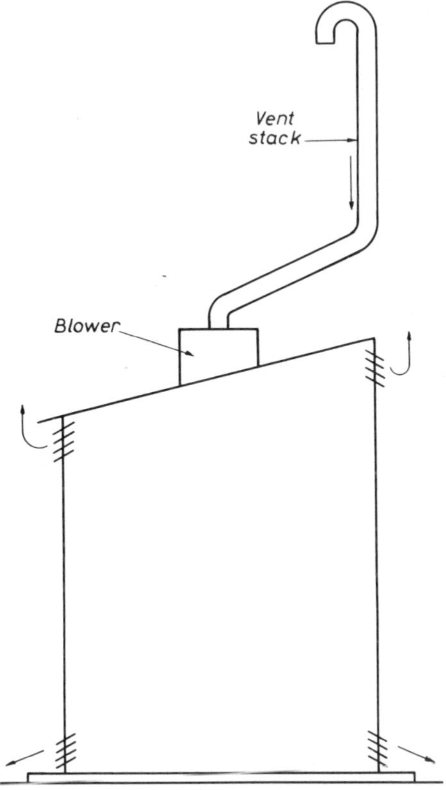

Fig. 9.4 — Analyser house ventilation.

9.3 FIRE HAZARDS

The main properties of liquids that are used to characterize the amount of fire
hazard they represent are as follows.

Flash point: the temperature at which a liquid will give off sufficient vapour
to form an ignitable mixture with air. The source of ignition may be a flame or a
spark. It is, at least under ideal conditions, equal to the temperature at which the
vapour pressure of the liquid is equal to the lower flammable limit of concentration.

Flammable limits: the lower (LFL) and higher (HFL) limits of concentration,
usually expressed as % v/v and reported for atmospheric pressure, are the lowest
and highest concentrations of the gas or vapour in air that can be ignited.

Autoignition temperature: the lowest temperature at which a fluid will ignite
spontaneously (measured as the temperature of a hot-plate on which the liquid

ignites when added in drops).

Table 9.1

Properties of some flammable fluids

Substance	LFL % v/v	HFL % v/v	Flash point °C	Autoignition temperature °C
Hydrogen	4.0	74.2		585
Methane	5.2	14.5	− 188	538
Ethane	3.0	15.0		515
Ethylene	3.0	34	−236	449
Propane	2.1	9.5	−104	468
Propylene	2.0	11.1	−108	410
n-Butane	1.9	8.5	− 60	405
Isobutane	1.8	8.5		462
But-1-ene	1.6	10.0	− 80	385
But-2-ene	1.7	9.7	− 73	324
Isobutane	1.7	9.0	− 76	465
Acetylene	2.5	82	− 18	299
n-Pentane	1.4	8.0	− 49	309
Isopentane	1.3	7.6		420
n-Hexane	1.2	5.5	− 22	240
n-Heptane	1.05	6.7	− 4	222
n-Octane	0.84	4.5	13	220
n-Nonane	0.8	2.9	30	206
Benzene	1.3	8.0	− 11	562
Toluene	1.2	7.1	4	508
o-Xylene	1.0	6.0	46	464
m-Xylene	1.1	7.0	29	528
p-Xylene	1.1	7.0	39	
Naphtha, hi-flash	0.9–1.3	6.0	38–43	482–510
Naphtha, solvent	1.1	6.0	38–43	232
Gasoline	1.2	7.6	− 43	257
Aviation gasoline	0.7	8	− 46	440
Kerosene	0.7	5	38–74	229
Gas oil	6.0	13.5	65	340
Methanol	7.3	36.5	11	463
Ethanol	3.3	19	13	422
Diethyl ether	1.7	48	− 47	180
Acetone	2.6	13	− 18	538
Methyl ethyl ketone	1.8	11.5	− 6	515
Ammonia	15	28		649
Carbon monoxide	12.5	74.2		609
Natural gas	3.8–6.5	13–17		482–632

Table 9.1 lists values of these properties for some common liquids and gases [1–4]. It should be noted, however, that the values should be used only to give guidance, because (*a*) the values determined and reported are dependent on the precise conditions used for the test, (*b*) the values determined for a particular material depend critically on the concentration of impurities in it, and (*c*) the flammable limits depend on the temperature of the gas and on the amount

of moisture present (increasing temperature and pressure widen the limits of flammability, especially downwards).

The (U.S.) National Fire Protection Association (NFPA) classifies a liquid as *flammable* if it has a flash point below 60°C, and as *combustible* if the flash point is greater than 60°C. If the flash point is above 93°C, the liquid is said to be not very hazardous.

The limits of flammability for a mixture of flammable gases are given by

$$\text{LFL} = \frac{100}{\Sigma(C_i/L_i)}\% \qquad\qquad \text{HFL} = \frac{100}{\Sigma(C_i/H_i)}\%$$

where C_i is the concentration (% v/v) of component i, and L_i and H_i are its lower and upper limits of flammability.

If conditions are suitable for combustion, a hot surface can cause spontaneous combustion if its temperature is higher than the autoignition temperature of the gas or vapour. The maximum surface temperature of equipment in a hazardous area should not be allowed to exceed 80% of the appropriate auto-ignition temperature.

Electrical Equipment

The installation of electrical equipment in a situation where a flammable atmosphere is, or could be, present, causes a risk of fire or explosion. In the U.K., the legal requirements are [5] that "all conductors and apparatus exposed to the weather, wet, corrosion, flammable surroundings or explosive atmosphere, or used in any special process shall be so constructed or protected, and such precautions shall be taken as may be necessary adequately to prevent damage in view of such exposure or use". The I.E.E. Wiring Regulations [6] give details of how this is expected to be done, and list some of the appropriate British Standards.

The most important standard is BS 5345 [7]. This Standard, which is relatively new and not yet published *in toto*, should be used for all new installations. Many analysers are still made to the older standards under which they were originally certified. New analysers will probably be certified against the newer standards.

Hazardous areas are classified as follows:

Zone 0: in which an explosive gas-air mixture is continuously present, or present for long periods.

Zone 1: in which an explosive gas-air mixture is likely to occur in normal operation.

Zone 2: in which an explosive gas-air mixture is not likely to occur in normal operation, and if it occurs will exist for only a short time.

An area that is not classified Zone 0, 1 or 2 is deemed to be a non-hazardous or safe area.

It should be noted that *Zones* were previously referred to as *Divisions*, and they are still so denoted in some parts of the world, especially America.

Electrical apparatus for use in hazardous areas must be protected in some way. The types of protection suitable for use in the various Zones are listed in Table 9.2.

Table 9.2

Zone	Protection	BS 5345 Part No.	BS 5501 Part No.	BS4683 Part No.
0	'ia' intrinsic safety	4	7	
	's' (specifically certified for use in Zone 0)	8		
1	'd' flameproof enclosure	3	5	2
	'ib' intrinsic safety	4	7	
	'p' pressurized	5	3	
	'e' increased safety	6	6	4
	's' special protection	8		
	also, any protection suitable for use in Zone 0			
2	'N' or 'n' non-sparking	7		3
	'o' oil immersion	9	2	
	'q' powder filling	9	4	
	also, any protection suitable for use in Zones 0 or 1			

BS 5345 gives codes of practice, BS 5501 gives requirements for construction and testing, and BS 4683 gives specifications for electrical apparatus.

In addition to the class of protection, the maximum surface temperature of the apparatus will also be of relevance to the choice of equipment for use in hazardous areas [8]. The temperature classes, and their respective maximum surface temperatures, in °C, are as follows: T1 – 450, T2 – 300, T3 – 200, T4 – 135, T5 – 100, T6 – 85.

Intrinsically safe apparatus (type of protection 'i') [9, 10]. Intrinsic safety is a protection technique based upon the restriction of electrical energy within apparatus and of interconnecting wiring, exposed to a potentially explosive atmosphere, to a level below that which can cause ignition by either sparking or heating effects. It is necessary to ensure that all interconnected apparatus, even if not exposed to the hazardous atmosphere, is also suitably constructed.

In the U.K., apparatus can be certified to be intrinsically safe by the British Approval Service for Electrical Equipment in Flammable Atmospheres (BASEEFA [11]), if it is manufactured and tested according to [12] or [12a]. In the USA, Underwriters' Laboratories, Inc. (UL) and Factory Mutual (FM) certify certain equipment to be safe according to the requirements of the various applicable NFPA and NEMA standards and codes [13]. In Germany, the Physikalische Technische Bundesanstalt (PTB) certify certain equipment according to the requirements of the Verband Deutscher Elecktrotechniker (VDE).

Guidance on the selection and installation of intrinsically safe apparatus is given in [9] and [10]. I.S. enclosures should be adequately identified. I.S. cables are normally coloured blue for easy recognition.

Flameproof enclosures (type of protection 'd') [10, 14]. A flameproof enclosure for electrical apparatus is one that will withstand an internal explosion of the flammable gas or vapour that may enter it, without suffering damage and without communicating the internal flammation to the external flammable gas or vapour for which the enclosure is designed, through any joints or structural openings in the enclosure.

Equipment which is capable of igniting a hazardous atmosphere, either by sparking or because of the presence of a hot surface or a flame, must normally be installed inside a flameproof enclosure if it is to be used inside a hazardous area, unless some form of special protection can be devised.

The standards for flameproof enclosures are [15] and [16], the certifying authority is again BASEEFA, and guidance on selection and installation is given in [10] and [14].

Segregation. If electrical apparatus is isolated by being placed outside the area of risk, or in an adequately isolated compartment, or separated by an impermeable barrier from exposure to a flammable concentration of gas or vapour, it is said to be segregated [17]. Where practicable, switch and control equipment should be installed outside any hazardous area, either in the open air, or in a special sealed room or compartment.

Expensive flameproof motors may not be needed if it is possible to pass a shaft from the motor through a suitable seal from the safe area to the hazardous one. This is a useful technique for use with fans that provide the suction in extraction ducting.

Ventilation [17]. Ventilation is the single most important means for prevention of formation of hazardous concentrations of gases or vapours, and hence of the risks associated with them. In some circumstances, good ventilation will convert a Zone 0 area into a Zone 1 area, and hence allow (flameproof) electrical equipment to be installed. More usually, it will allow an analyser house to be reclassified from Zone 1 to Zone 2, or from Zone 1 or 2 to *safe area*.

If a flammable fluid is brought into unventilated enclosed premises, the area is classified as Zone 1, whether or not the fluid is normally released, but if there is adequate ventilation, the classification may be reduced.

The greatest hazard arising from a source of flammable gas or vapour extends over the area in which the concentration can exceed the LFL. The extent of the area will depend on the effects of air flow, obstructions, and on whether the gas or vapour is less or more dense than air. A dense vapour tends to drop and diffuse outwards at ground level; it is normal practice to assume a radius of hazard of 15 m at the elevation of the source, but of 30 m at ground level and 1 m above it. Gases that are less dense than air diffuse rapidly upwards, so the radius of hazard at the elevation of the source is rather less (3–10 m) and there should be little hazard at ground level.

Under normal operating conditions, ventilation should be such that the atmosphere in the analyser house has an almost undetectable trace of flammable gas in it – certainly less than 5% of the LFL. Mechanical ventilation is normally necessary. Roof-mounted blowers with diffusers and air ducts to the space between analysers give good distribution of fresh air. Inlet air should be taken from a safe area, and gases should be discharged at an exposed position in the open air, away from features where 'pocketing' of the gases could occur, and remote from any source of fresh air used for ventilation. A typical ventilation-air flow-rate is sufficient to give 20–100 air changes/hr in the analyser house.

In order to reduce the entry of hazardous gases into the analyser house, it is usual for the air within the house to be maintained at a pressure of >5 mmH$_2$O.

The ventilation system must be capable of keeping the flammable gas concentration below 20% of the LFL, or at worst, below 40% of the LFL, even in the event of sample-line breakage. In order to minimize the amount of hazardous sample entering the analyser house, especially in the event of a sample leak, it is advisable to ensure that any fast by-pass sample-conditioning systems are located outside the analyser house. Leakages inside the house should be minimized by use of flow limiters or excess-flow check valves.

Special attention must be given to the possibility of failure of the ventilation system, or of the analysers, pipework etc. The following alarms and interlocks provide protection against such failures.

(i)   *Ventilation-air flow-failure.* Failure of the ventilation must be detected and annunciated if it is necessary to keep the concentrations of flammable gases at a level that is consistent with the classification of the area.

(ii) *Air pressurization failure.*

(iii) *Door open.* If a door is open, the ventilation and pressurization will be disturbed, unless there is an air-lock. Thus, 'Door Open' alarm and interlock switches should be fitted. However, the interlock switch should not cause shut-down of the analysers if the door is left open for less than five minutes, since the occurrence of a hazardous condition in such a short time would be unlikely, and it would be undesirable for the analysers to become unavailable if the doors are left open only for very short periods.

(iv) *Flammable gas alarm*. The concentration of flammable gas in the analyser house air should be less than 20% of the LFL if the analyser house is a safe area. To ensure that this is so, combustible-gas detectors should be installed; the number required depends on the size of the analyser house, on the number and location of potential gas sources, and on the distribution of the air flow. Commonly, one detector is fitted for every 10 m^2 of floor space, provided that there are no regions of poor ventilation. Detectors are sometimes fitted near the roof and the floor to detect concentrations of light and heavy gases. They should be set to operate an alarm if the flammable gas concentration exceeds 20% of the LFL, and to cut off the electrical supply to the analysers if it exceeds 40% of the LFL.

(v) *Emergency shut-down*. A covered weatherproof push-button should be fitted outside the analyser house near to each door. When this is depressed, the power to all equipment in the analyser house should be isolated and an alarm annunciator should operate in the plant control room. This could be called a fire-alarm button, in which case the condition should be annunciated in the fire station. To avoid a false shut-down, no other switch should be situated near to this push-button.

Before any shut-down, prior annunciation should be provided where possible. However, it is not necessary to give prior warning of a shut-down condition if the time interval is too short for any action to be taken.

Table 9.3 lists the possible alarms and interlocks, and gives their uses.

Table 9.3

Use of alarms and interlocks

Analyser house area classification change	Ventilation alarm and interlock switches required
Zone 1 to Zone 2	Low-flow alarm and door-open alarm. Low-flow, door-open and timer switches in interlock circuits of non-Zone 1 equipment.
Zone 2 to Safe Area	Low-flow alarm and door-open alarm. Low-flow, door open and timer switches in interlock circuits of non-explosion-proof equipment.
Zone 1 to Safe Area	Low-flow, low-pressure, door-open and sometimes high flammable-gas concentration alarms. Low-flow, low-pressure, door-open and sometimes high flammable-gas concentration and timer switches in interlock circuits of equipment not suitable for Zone 1 area. Equipment made safe by purging must be isolated if purge-gas is insufficient.

The following calculation illustrates the procedure for determining the number of air changes required to reduce the flammable gas concentration to 40%

of the LFL in the event of breakage of a sample line within an analyser house.

> Sample – ethylene vapour, LFL 3.0% v/v
> Maximum flow-rate – 51 m³/hr
> 40% of the LFL = 1.2% v/v
> Air flow-rate required = 51 × (100/1.2) ∼ 4250 m³/hr
> If the volume of the enclosure = 34 m³
> Number of air changes required per hour = 4250/34 = 125

This is a rather higher ventilation rate than normal, so in this instance, it would be desirable to take some action to reduce the flow-rate of ethylene inside the analyser house.

Purging and pressurization (type of protection 'p' [17]). These may be regarded as a special local application of ventilation. The use of purging and pressurization requires special enclosures which should be made as air-tight as possible. Generally, before use, such an enclosure must be purged by at least five volumes of purge gas before power is switched on, in order to ensure that any hazardous gas trapped inside has been removed or diluted to 20% of the LFL or less. Provided that no flammable gas can be released inside it, the enclosure is then pressurized to slightly above atmospheric pressure (5 mmH₂O) so that any leakage will be from the inside of the enclosure to the outside, ensuring that a surrounding flammable atmosphere cannot reach the electrical equipment contained inside. If flammable gas is released within the enclosure, purging must be continued, though the rate can be decreased from the initial high value.

Purging and pressurization are covered by the European CENELEC standard EN50 016 (1977), published in Britain as BS 5501: Part 3, which applies to enclosures *without* an internal source of release of flammable gas and *without* personnel within the enclosure. The code of practice BS 5345: Part 5, which is expected to be published at about the same time as this book, will apply to enclosures *with* an internal source of gas release such as analysers. Equipment is already being certified by BASEEFA for such applications and is being issued with Ex'p' certificates.

Three types of enclosures may be distinguished.

(*a*) Enclosures without an internal source of gas release, purged with air or inert gas. After the initial purge, the flow of gas is stopped and a positive pressure is maintained. This is known as Leakage Compensation – LCA (air) or LCG (inert gas).

(*b*) Enclosures with an internal source of gas release, purged with compressed air. The initial purge must be continued, perhaps at a lower rate, but sufficiently to ensure that the gases are diluted at least to 20% of the LFL. This is known as Continuous Dilution – CDA.

(*c*) Enclosures with an internal source of gas release purged with inert gas. The initial purge will remove the oxygen from within the enclosure, after which a positive pressure is maintained; the purge rate can be reduced to a low value. Either the LCG or CDG methods can be used.

Applications such as (b) and (c) normally create a Zone 1 area within the enclosure, so it is advisable for power to be shut off automatically if the supply of purge and pressurizing gas fails. In the case of (b), it is important to remember that the area surrounding the purge gas exhaust will be classed as hazardous. If it takes some time for the electrical energy to decay after the power supply has been disconnected, there must be a delayed unlocking device or a notice stating that the enclosure must not be opened until the required period has elapsed.

If the continuous running of an analyser is important, e.g. for process control, an emergency alternative supply of purge gas should be available for use if the main source fails. An annunciator to give prior warning of purge-failure should be fitted, so that action can be taken before the actual failure occurs.

It is often considered that static pressurization is adequate for non-explosion-proof analysers to be used in a Zone 2 area, and that purging is only necessary in Zone 1 areas. However, it is good practice to use purging wherever possible.

A purge and pressurization unit is made by GEC-Elliot for use with their analysers. The unit gives an initial timed purge, then maintains a pressure of air in the instrument. Before the analyser is connected to the electrical supply, any trapped flammable gases are purged by means of $\sim$ 5 air changes. The purge rate, selected to take account of the probable rate of inflow of flammable gas, varies from 10 to 100 changes/hr. With a typical flow of 20 changes/hr, a purge time of 15 min will be required to achieve the necessary 5 air changes. Once the purge period has elapsed, the analyser is pressurized and can begin to operate.

Systems designed to receive Ex'p' certificates are made up by the company Expo Safety Systems Ltd. These have component approval, and in some cases (if the schedule of limitations issued by the relevant European Test Houses is complied with) will be covered by a full certificate.

Special protection (type of protection 's') [18]. The concept of special protection is one that allows some flexibility, in that it permits certification of electrical equipment which does not conform to the requirements for other types of protection, but which can be shown (e.g. by test) to be suitable for use in the specified hazardous area(s).

An example of an analyser that can be specially protected is a gas analyser containing a hot-wire or catalytic detector element; such elements are ignition sources in normal operation. If they are made safe by fitting flame arresters, such analysers may be certified 's'.

Apparatus for Zone 2 Areas. The requirements for instruments operating in Zone 2 areas are not so stringent as for Zones 0 and 1. For example, protection type 'N' [19] is suitable. Guidance on installation and selection of equipment is available [20,21]. However, much of what applies to Zones 0 and 1 also applies to Zone 2, and it is recommended that apparatus constructed to the more rigorous standards required for Zones 0 and 1 be used whenever possible in Zone 2 also.

Isolation of the electrical supply to the analyser in the event of purge-gas failure may not be necessary in a Zone 2 area.

Servicing. If the analyser enclosure is a safe area of adequate size, the equipment may be serviced on site with the power switched on, but if it is not a safe area, the equipment must be serviced elsewhere if it has to be opened while the power is on. In some circumstances, the local safety inspector may measure the gases in the area, and, if it is safe, issue a permit for the equipment to be serviced on site.

For optimum safety, there should be no electrical supplies or signals in the analyser house that are not intrinsically safe and which are not obtained from the analyser house supply.

9.4 COMBUSTIBLE-GAS MONITORS

The detector cell of one type of explosimeter or combustible-gas monitor consists of a resistance thermometer, the filament of which is heated by an electric current and also by the combustion of hydrocarbons at the surface of the hot filament. The heat resulting from combustion on the hot filament is proportional to the concentration of hydrocarbon vapour present in the air, so the temperature change, as measured by the resistance of the thermometer, is a direct measure of the concentration of the hydrocarbon vapour in the air. The cell is connected in a Wheatstone bridge which also has a sealed reference cell containing air and two resistors. The out-of-balance voltage is amplified so that it can operate an indicating meter and a switch which, in turn, operates an alarm system.

Earlier detector cells consisted of a platinum wire on the surface of which catalytic combustion occurred at about $1000°C$. The life of the element was rather short, about 3–6 months. To obtain an increased life, the functions of temperature detection and of catalyst for combustion may be performed by two materials. The platinum resistance thermometer is coated with a bead consisting, typically, of a mixture of palladium and thorium oxide which, for a similar sensitivity, only has to work at a temperature of the order of $500°C$, resulting in a considerably longer life, with negligible damage to the platinum wire.

Palladium–platinum alloys are used to catalyse the oxidation of hydrocarbons. The hydrocarbon molecules are adsorbed on the surface of the catalyst. Hydrogen atoms are held at certain preferred active positions by attractive forces between the metal and the hydrogen atoms. This weakens the chemical bonds which bind the hydrogen atoms to the molecules, thus making the adsorbed molecules more reactive to oxygen.

The net effect of this adsorption/oxidation process is that heat is generated, the amount depending on the number and size of the hydrogen-containing molecules adsorbed onto the catalyst surface. The catalyst reduces the temperature at which the hydrocarbon oxidizes. For the hydrocarbon to ignite, the sample must contain a sufficient concentration of oxygen.

Bacharach make a very small platinum–iridium alloy hot-wire element (maintained at about 65°C) which is said to have a life of 6–36 months. It is enclosed by a suitable filter to reduce the effect of filament poisons such as silicones. Noise, which is mainly thermal, limits the minimum detectable concentration.

Decomposition of the sample at the hot filament can result in an apparently safe condition. If this can occur, it is advisable to under-run the filament.

The low-temperature catalytic-bead type of detector is poisoned by lead octane-improvement additives; therefore, when these can be present in the atmosphere, the high-temperature platinum-wire detector should be used.

Sensor-element poisons react with part of the surface or coat part of the surface and thus reduce the available active area.

As an alternative to resistance thermometers some manufacturers use thermocouples. The active and reference thermocouples are connected with the signals opposing.

The alarm switch should be set to operate at about 20% LFL and the trip or shut-down switch should be set to operate at 40–60% LFL, depending on the location and elevation of the detector. These levels might be selected for a Zone 2 area, but a lower level might be preferable if the area is (or is to be) classified as safe.

Detectors should be mounted high, say 0·5 m from a roof, to detect flammable gases lighter than air, and low, say 0·5 m from floor level, to detect flammable gases heavier than air.

A flammable concentration of gases can exist, however, at other elevations and the best location of detectors in a building is always difficult to determine. If the building is well ventilated and the air is well circulated then the detector should be located near to the source of the gas, but this still may not be the location of the accumulation of the greatest concentration.

Various companies make flame ionization monitors which can be used as explosimeters.

Flammable gas monitors, also known as explosimeters, are described in detail in Book 4 of this Series; Chapter 17 describes solid-state gas monitors, which are used to monitor flammable and toxic gases in air.

The sensitivity of these monitors is different for different gases. At one time it was thought that the relative sensitivities depended only on the ratio of concentrations at the lower flammable limits (LFL %), but it is now known that they also depend on the heat of combustion of the components and on the size of the components (especially for diffusion-type sensors). As an example, consider the response of a sensor calibrated for propane.

Gas	LFL, %v/v	$LFL_{gas}/LFL_{propane}$	Observed sensitivity (propane = 1.00)
Methane	5.0	5.0/2.1 = 2.4	1.81
Ethane	3.1	3.0/2.1 = 1.4	1.22
Propane	2.1	1.0	1.00
Butane	1.85	1.85/2.1 = 0.9	1.11
Pentane	1.4	1.4/2.1 = 0.67	0.84
Hexane	1.2	1.2/2.1 = 0.57	0.66

In order to reduce the effect of variations in wind direction, it is possible to locate several sensors around and downstream of the probable source. The sensors should have individual local amplifiers, each connected to a high-level signal selector; an averager should not be used. Sometimes an averager is used in preference to a multi-point scanner where it can be assumed that it is improbable that more than one source will produce an appreciable contamination of the air.

The Control Instruments Corporation makes a flammable gas detector which consists of a resistance thermometer that measures the temperature of a flame.

The I.S.T. (International Sensor Technology) hydrocarbon-gas sensor uses a solid-state sensor system, and there is no combustion, no temperature increase, and no requirement for oxygen in the operation. It requires calibration approximately every six months.

It is possible to monitor fairly high concentrations of ammonia in air such as those resulting from serious leaks in refrigeration plants, by the use of a flammable-gas detector without oxygen enrichment. The lower flammable limit is about 15% v/v of ammonia in air, so an explosimeter (which will probably be adjusted for a range of 0–12.5%) may be scaled to read 0–2% v/v, which is acceptable for leak detection. This has been successfully applied by Draeger, using the Crowcon Gaswarden.

9.5 OTHER HAZARDS

Three major hazards must be considered when examining the design of those components of analyser systems which are located in the field. The first is the fire hazard which can exist when flammable vapours are handled by equipment containing electrical components or other devices which can cause a spark or a hot surface. The second is the toxic hazard which, on the rare occasions which it is important, affects operators and not equipment and is often not easy to detect by normal physical means. The third is the rarer radioactive hazard which, like the toxic hazard, also affects operators and not normally the equipment.

9.5.1 Oxygen depletion

Wherever there is access for personnel, the concentration of oxygen in the atmosphere should not be less than 16% v/v and preferably not less than 18%

v/v. If there is any danger of oxygen depletion a monitor should be fitted. However, the toxicity of the gas which replaces the air could cause a hazard that is far greater than the depletion of oxygen itself.

In tank farms, it is necessary to measure the concentration of the unreduced oxygen in the blanket gas. The range of the analyser for this should be 0–5% v/v.

Another important oxygen measurement in tank farms which is not so commonly used is oxygen deficiency. This is important if the blanket gas, which is heavier than air, leaks out of the tanks and builds up at a low level around the tanks before it disperses.

9.5.2 Radiation

Ionizing radiation is relatively easy to detect, provided that the type of radiation likely to be present is known. The effects of over-exposure to radiation are not usually immediately detectable, which makes radiation doubly hazardous. Radiation protection is now an important independent discipline on which many texts are available, e.g. [22–25].

In the U.K., there are numerous regulations, codes of practice, recommendations, etc., regarding the handling of radioactive materials [22, 25–30]. The laws of countries other than the U.K. are briefly discussed in [22].

If it is intended to use a radioactive source on a plant, the Health and Safety Executive must be notified *before* installation of the equipment containing the source.

The *activity* of a radioactive source is measured in *curies* or in *becquerels* (the SI unit). A source with an activity of 1 Ci undergoes 3.7×10^{10} disintegrations per second.

$$1 \text{ Bq} = 1 \text{ disintegration per sec} = 27.03 \times 10^{-12} \text{ Ci} .$$

The *absorbed dose* of radiation is the energy imparted to matter by ionizing particles, per unit mass of irradiated material at the place of interest. It is measured in terms of the *rad* or the *gray* (the SI unit).

$$1 \text{ rad} = 10^{-2} \text{ J/kg}$$
$$1 \text{ Gy} = 100 \text{ rad} = 1 \text{ J/kg}$$

The *dose rate* may be expressed in rad/hr, etc.

The magnitude of any biological effect from absorption of ionizing radiation depends not only on the absorbed dose or dose rate, but also on the type of radiation and on some other factors. For this reason, the *dose equivalent* had to be introduced. The dose equivalent, units *rem* (rad-equivalent-man) or *sievert* (the SI unit) is numerically equal to the absorbed dose multiplied by the quality factor (QF), the dose distribution factor (DF) and by other modifying factors.

That is

$$A = D \times QF \times DF \times \ldots\ldots$$

If D is measured in rad, A will be in rem; if D is in Gy, A will be in Sv (1 Sv = 100 rem). Table 9.4 gives the values of QF for some types of radiation.

Table 9.4

Type of radiation	QF
X, γ, β ($E_{max} > 30$ keV)	1.0
β ($E_{max} < 30$ keV)	1.7
fast neutrons and protons to 10 MeV	10
α-radiation from natural radioactive decay	10
Heavy recoil nuclei	20

Dose equivalents are additive but absorbed doses are not. If the dose rate is known, the total dose is found as the product of the dose rate and the time of exposure.

The recommended maximum permissible dose for adults exposed to radiation in the course of their work is 5 rem/yr. Members of the public should not receive more than 0.5 rem/yr [26].

Typical levels of leakage from the shielded sources contained in some on-line analysers are reported to be less than 0.1 mrem/hr at distances between 10 and 100 cm.

Any person working with radioactive material must normally be 'designated', unless it can be shown that his dose is most unlikely to exceed 30% of the maximum permissible. Such a person must have medical examinations, and must wear some type of monitoring device (dosimeter). Normally, this will consist of a film badge [31], but other possibilities include thermoluminescent dosimeters, ionization chamber dosimeters or Geiger-counter ratemeters [22–25, 32]. Detailed information on film badges may be obtained from the National Radiological Protection Board, Harwell, Didcot, OX11 0RQ, U.K.

Non-ionizing radiation
There has been little legislative interest in safety precautions for non-ionizing radiation, but the following points should be borne in mind. *Ultraviolet* light can cause eye damage [25], and may produce excessive ozone in enclosed spaces. Intense *infrared* radiation may also cause eye damage (cataracts). *Microwave* radiation causes intense heating inside tissue (as in cooking in microwave ovens), so any leakage is potentially dangerous. As yet, little is known about safe limits, so no conclusions are possible. *Laser beams* are generally recognized to be hazardous: some guidance about precautions is available [33].

9.5.3 Toxic hazards

Toxicity is the relative poisonous effect, and a poison is a chemical that causes damage to the body. The severity of a toxic effect depends on the nature of the toxic substance, the quantity of the substance, the duration of exposure, the affinity of the body and the sensitivity of the body. Poisons may be absorbed through the skin, the mouth and gastrointestinal tract, or the lungs.

The Threshold Limit Value (TLV) of a toxic gas, vapour or dust is a time-weighted average concentration to which workers can be exposed for 8 hours a day, 5 days a week. The maximum allowable concentration (MAC) of a substance is a concentration that should not be exceeded at any time during the working day; MAC values are larger than the corresponding TLVs. The recommended TLVs are based on decades of industrial experience; they are listed in safety books [2, 25] and in special publications [34, 35].

Some guidance on the use of toxic materials is given in a Health and Safety Executive Guidance Note [36] and in a recent text [37].

Most hydrocarbons are between slightly and moderately toxic, with a typical TLV of 200–1000 ml/m^3. Hydrogen sulphide is very toxic, with a TLV of 10 ml/m^3; when exposure to hydrogen sulphide is only intermittent, 15 ml/m^3 can be regarded as the maximum safe concentration, and a personnel-protection alarm can be set to operate at this level.

In enclosed spaces, toxic gases should normally be diluted to a concentration lower than the TLV for safe exposure of the operators.

Table 9.5

TLV and MAC values for some substances

Substance	TLV ml/m^3	MAC ml/m^3
Ammonia	25	35
Benzene	10	10
Carbon monoxide	50	(400)
Chlorine	1	3
Octane	300	375
Hydrogen cyanide – skin	10	10
Hydrogen sulphide	10	15
Xylene – skin	100	150

In the U.S.A., the Underwriters' Laboratories group gases in several categories according to the toxicity of the gas. Category 1 gases are the most toxic and cause death if a person is exposed to about 1% concentration of the gas for about 5 min. Category 6 gases are the least toxic; exposure to a 20% v/v concentration for 2 hr should cause no signs of illness.

If the analyser enclosure has a natural ventilation of V m^3/hr and there is a concentration of toxic substance of C ml/m^3, then the ventilation must be increased to V_2 m^3/hr to decrease the concentration of the toxic substance to the acceptable TLV. The ventilation should further be increased to include a safety factor (K) to cater for the effects of bad distribution of dilution air, so that the TLV is not exceeded:

$$V_2 = \frac{C \times V \times K}{\text{TLV}} \, \text{m}^3/\text{hr}$$

If a toxic vapour is produced at a rate of Q m^3/hr then

$$V_2 = \frac{Q \times K}{\text{TLV}} \text{m}^3/\text{hr}$$

No general poison-detecting instrument is available; however, if the nature of the poison is known, its concentration can be measured if the level is sufficient. If the presence of a poison is suspected, it is advisable to be observant and to take adequate personal precautions such as minimizing the duration of exposure, wearing a gas mask, wearing goggles and washing thoroughly. The capacity of a gas mask is not unlimited. For typical toxic vapours or gases, it should not be used in a toxic gas concentration of 2% v/v for more than about 20 min.

The effects of *similar* toxic substances are additive. The TLV of such a mixture is T ppm and is given by

$$\frac{C_{\text{tot}}}{T} = \Sigma \frac{C_i}{T_i}$$

where C_i and T_i are the concentration and TLV (both in ppm v/v) for the ith component, and C_{tot} is the total concentration. If $C_{\text{tot}}/T > 1$, the threshold has been exceeded.

If the components in the mixture are *dissimilar*, their effect may be independent, or additive, but may also be greater than expected from additivity (synergic effect) or less (antagonistic effect).

Examples of toxic gas monitors
The concentration of chlorine can be determined by a conductivity analyser which compares the conductivity of the fluid before and after the reaction:

$$\text{AgNO}_3 + \text{Cl}_2 + \text{H}_2\text{O} \longrightarrow \text{AgCl} + \text{HClO} + \text{HNO}_3 \, .$$

Chlorine-leak detectors are available from Beckman, Du Pont, Fischer and Porter, Wallace & Tiernan, and Wösthoff.

The MSA carbon monoxide alarm works on the principle of the selective oxidation of CO by Hopcalite and measurement of the temperature rise due to the heat generated by the exothermic reaction. In order to minimize the temperature error, the temperature difference is measured between the active Hopcalite bed and an inactive bed. Water poisons the Hopcalite, so the active and inactive beds are controlled at a temperature slightly greater than 100°C. The range is 0–500 ml/m^3 with an error of ±1% FSD. The 90% response time is 45 sec. The cell (Fig. 9.5) is designed to maintain the sample temperatures in both halves of the cell as near to equality as possible.

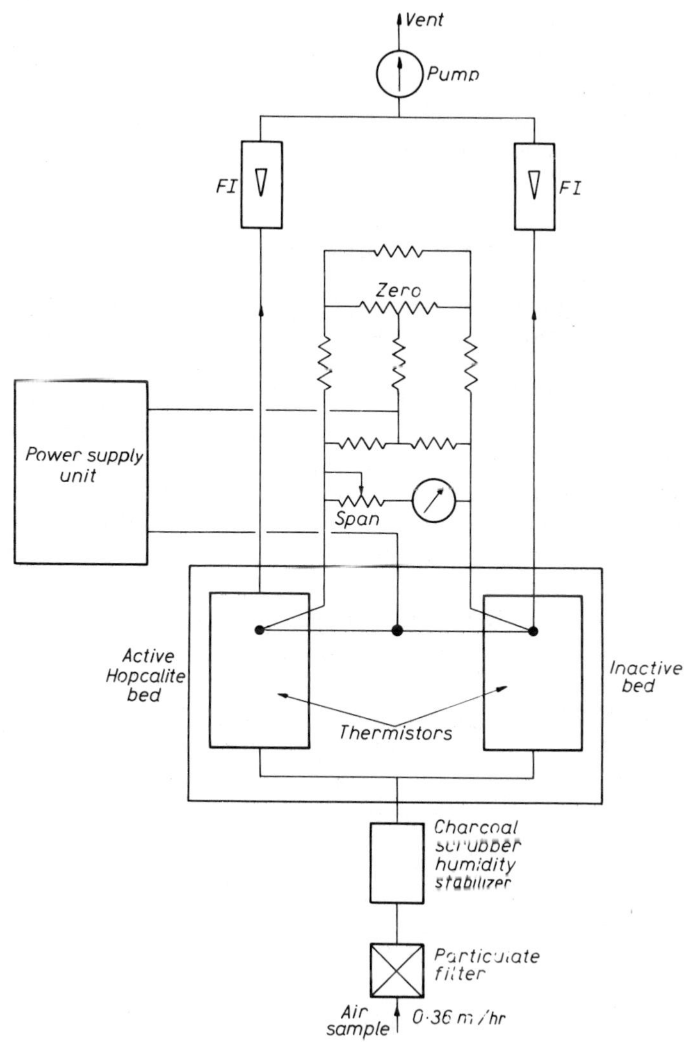

Fig. 9.5 — The MSA carbon monoxide alarm.

The Bacharach Instrument Company makes the Johnson–Williams toxic-gas alarm, which is based on the increased conductivity of distilled water when certain gases such as sulphur dioxide and ammonia dissolve in it. Hydrogen sulphide is detected by passing it through a heated reactor where it is converted into sulphur dioxide. The concentration of nitrogen dioxide, chlorine and hydrogen chloride can also be measured by the instrument.

Solid-state analysers are being used increasingly for measurement of the concentration of a wide variety of gases in air, for plant and personnel protection. They are very popular for the measurement of H_2S and for this application monitors using metal oxide sensor elements are available from Bacharach, the Ecolometrics Division of General Monitors, and International Sensor Technology. Similar monitors are also available for the measurement of the concentrations of ammonia, chlorine, sulphur dioxide, carbon monoxide and hydrocarbons in air.

9.6 PRESENTATION ON THE CONTROL PANEL

One of the best ways of presenting on a panel the location of a toxic, flammable or other hazard is a plastic lay-out diagram which shows the locations of the major equipment of a process plant, especially those parts which could be the source of a hazard or failure. On the diagram are located a number of small lamps or light-emitting diodes (LEDs) which represent the location of each of the hazardous-gas detectors. A lamp (or LED) is illuminated when the corresponding hazard has been detected. Adjacent to the diagram, a wind-direction indicator should be installed. The diagram lamps should not replace those of the annunciator system and the monitors.

The overall systems must be as reliable as possible and should be capable of being tested as completely and as simply as possible.

Some diagram lamps are provided to indicate the danger condition as well as the warning condition (Fig. 9.6).

The I.S.T. monitor has small indicator lamps for the following functions. The function names used have not, unfortunately, been adopted universally.

Table 9.6

I.S.T. Functions

Name	
WARN	Warning or pre-alarm level, also named 'Caution' by Teledyne. This should *not* be called a low alarm.
ALARM	Danger or action alarm level.
ACTIVE	Power supply on.
FAULT	System failure. This should include line failure, open circuit and short circuit.

The active and fault functions are sometimes combined to operate a failure alarm.

The level of hazardous gas concentration should be indicated on the front of each individual monitor or by a common meter and a selector switch. Several modern instruments have a digital read-out.

It is possible to feed the output signals from the individual monitors to a summing card. The output of this summing card is then a measure of the total amount of toxic gas released. This could then operate an area alarm and, as a result, the operator could look at each monitor in order to investigate the source of the leak.

Different ranges are needed, according to the purpose of the monitor, which may be for personnel protection, plant protection, or leak detection.

In order to maximize the availability of these essential monitors, they are normally operated from a battery supply: an a.c. supply system may be used if it can be proved to be sufficiently reliable.

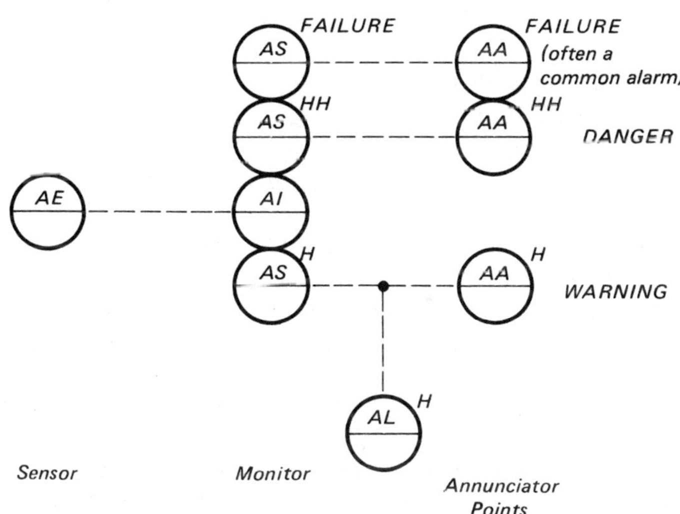

Fig. 9.6

REFERENCES

[1]   The Associated Factory Mutual Fire Insurance Companies, *Ind. Eng. Chem.*, 1940, **32**, 880.

[2]   N. I. Sax, *Dangerous Properties of Industrial Materials*, 4th Ed., Van Nostrand Reinhold, New York, 1975.

[3]   *Fire-hazard Properties of Flammable Liquids, Gases and Volatile Solids*, NFPA, Vol. 1, No. 325.

[4]   *Hazardous Chemicals Data,* NFPA Vol. 2, No. 49.

[5]   *Electricity (Factories Act) Special Regulations* 1908 and 1944, S, R and O 1908 No. 1312 and 1944 No. 739.

[6]   Institution of Electrical Engineers, *Regulations for the Electrical Equipment of Buildings,* 14th Ed. with amendments, I.E.E.E., London, 1976.

[7]   *Selection, Installation and Maintenance of Electrical Apparatus for Use in Potentially Explosive Atmospheres,* BS 5345. Part 1, Basic requirements; Part 2*, Classification of hazardous areas; Part 3, Type of protection 'd' – flameproof enclosure; Part 4, Type of protection 'i' – intrinsically safe apparatus; Part 5*, Type of protection 'p' – pressurization and continuous dilution; Part 6, Type of protection 'e' – increased safety; Part 7, Type of protection 'N'; Part 8, Type of protection 's' – special protection; Part 9*, Types of protection 'o' and 'q' – oil-immersed and sand-filled apparatus; Part 10*, Apparatus for use with combustible dusts; Part 11*, Specific industry applications; Part 12*, The use of gas detectors.

[8]   BS 5345: Part 1: 1976, BS 4683: Part 1: 1971.

[9]   BS 5345: Part 4: 1977.

[10]   BS CP 1003: Part 1: 1967.

[11]   Department of Trade and Industry, BASEEFA, Harpur Hill, Buxton, Derbyshire.

[12]   *Intrinsically Safe Electrical Apparatus and Circuits,* BS 1259: 1958.

[12a]   *BASEEFA Certification Standard* SFA 3012: 1972.

[13]   *U.S. Standards and Codes.* NFPA: 70-1981 – National Electric Code, NFPA: 493-1978 – Intrinsically safe apparatus in Division 1 hazardous locations, NFPA: 496-1974 – Purged and pressurized enclosures for electrical equipment, NFPA: 497-1975 – Classification of Class I hazardous locations for electrical installations in chemical plants, NEMA: ICS6-1978 – Enclosures for industrial controls, UL25-1973 – Standard for meters for flammable liquids and LP gas, UL698-1973 (ANSI: C33. 30-1973) – Standard for industrial control of equipment for use in hazardous locations, Class I, Groups A, B, C and D and Class II, Groups E, F and G, UL913-1976 – Standard for intrinsically safe electrical circuits and equipment for use in hazardous locations.

[14]   BS 5345: Part 3: 1979.

[15]   *Electrical Apparatus for Explosive Atmospheres,* BS 4683: Part 2: 1971. Also, BS 229: 1957, which it replaces, but which still applies for apparatus certified under it.

[16]   *Electrical Apparatus for Potentially Explosive Atmospheres,* BS 5501: Part 5: 1977.

[17]   BS CP 1003: Part 2: 1966, BS 5501: Part 3, BS 5345: Part 5.

[18]   BS 5345: Part 8: 1980.

[19]   BS 5345: Part 7: 1979, BS 4683: Part 3: 1972.

[20]   BS CP 1003: Part 3: 1967.

[21]   *Guide to Electrical Equipment for Use in Division 2 Areas,* BS 4137: 1967.

[22]   A. Martin and S. A. Harrison, *An Introduction to Radiation Protection,* 2nd Ed., Chapman & Hall, London, 1979.

[23]   H. F. Henry, *Fundamentals of Radiation Protection,* Wiley, New York, 1969.

[24]   K. Z. Morgan, and J. E. Turner, *Principles of Radiation Protection,* Wiley, New York, 1967.

[25]   L. Bretherick. *Hazards in the Chemical Laboratory,* 3rd Ed., Royal Society of Chemistry, London, 1981.

[26]   *Recommendations of the International Commission on Radiological Protection,* IRCP Publication 26, Pergamon Press, Oxford, 1977.

[27]   *Health and Safety at Work Act,* 1974, and regulations made and to be made under it. See also: Consultative Document — *Ionising Radiations, Proposals for Provisions on Radiological Protection,* HSC, London: HMSO, 1978.

[28]   Ministry of Labour, *Ionizing Radiation Precautions for Industrial Users, Safety, Health and Welfare Series* No. 13, HMSO, London, 1961.

[29]   *The Ionizing Radiations (Sealed Sources) Regulations,* 1969, (Statutory Instrument 1969 No. 808) (due to be repealed).

[30]   *Reports of the National Council on Radiation Protection and Measurements* (USA).

[31]   *Film Badges for Personnel Radiation Monitoring,* BS 3664: 1963.

[32]   *A.S.T.M. Methods* D-1671, D-1672, D-2568.

[33]   *A Guide for Control of Laser Hazards,* The American Conference of Governmental Industrial Hygienists, P.O. Box 1937, Cincinnati, Ohio 45201, USA, 1973; D. Sliney and M. Wolbarsht, *Safety with Lasers and Other Optical Sources,* Plenum, New York, 1980.

[34]   Health and Safety Executive, *Threshold Limit Values for 1979, Guidance Note* EH15/79, HMSO, London, 1980.

[35]   *Documentation of Threshold Limit Values for Substances in Workroom Air,* ACGIH, Cincinnati, Ohio, 1979.

[36]   Health and Safety Executive, *Toxic Substances: A Precautionary Policy, Guidance Note* EH18, HMSO, London, 1978.

[37]   C. F. Cullis and J. G. Firth, *Detection and Measurement of Hazardous Gases,* Heinemann, London, 1981.

*Not yet published (January, 1982).

CHAPTER 10

Signals and recording

The *output signal* from an analyser is normally recorded continuously by means of some form of chart recorder. The recording gives a continuous record of the physical or chemical quality being measured, and thus relates signal to time. The output signal can be in the form of a pneumatic signal in the pressure range 3–15 psig (0.2–1.0 kg/cm^2 g) or a d.c. electrical signal for long signal lines (e.g 4–20 mA, or a shorter span within this range), according to the make of process instrument. Potentiometric equipment gives a d.c. voltage signal, and if this is at the level of only a few mV, it is undesirable to have a long signal line, on account of its appreciable resistance and probable pick-up of noise.

Signal preamplification at the analyser is advisable if the signal level is low and subject to interference. The output impedance should also be sufficiently low. The use of high-frequency signals can cause problems. Coaxial cables, which are expensive, may be needed to decrease the cable capacitance.

The output signal may be presented in various forms. It can be recorded continuously as a function of time, as shown for the signal from a chromato-graphic analyser, in Fig. 10.1(*a*). A more economic method is *bargraph* presentation, (*b*) in Fig. 10.1. The chart drive is stopped when a component is being eluted, so that the normal peak is compressed into a line, the height of which is the same as the peak-height. A 'gating' system is used for this purpose. The paper moves between the appearance of gated peaks. The gate is opened and closed by departures from and return to the baseline signal. As the paper is advanced only a few mm between gates, the method saves a good deal of paper and records much information on a small length of chart. However, recording of the full chromatograph is useful for detection of column or gating failure, and is needed for correct setting of the gates and routine checks of performance.

In *trend signal recording,* one particular peak is selected and any change in the signal corresponding to it is updated at each analysis cycle, (*c*) in Fig. 10.1. Again the chart paper is advanced only a short distance between cycles, giving economy and rapid assessment of the changes in the signal (hence the term 'trend signal'). A separate memory/recorder unit is needed for each component to be displayed as a trend.

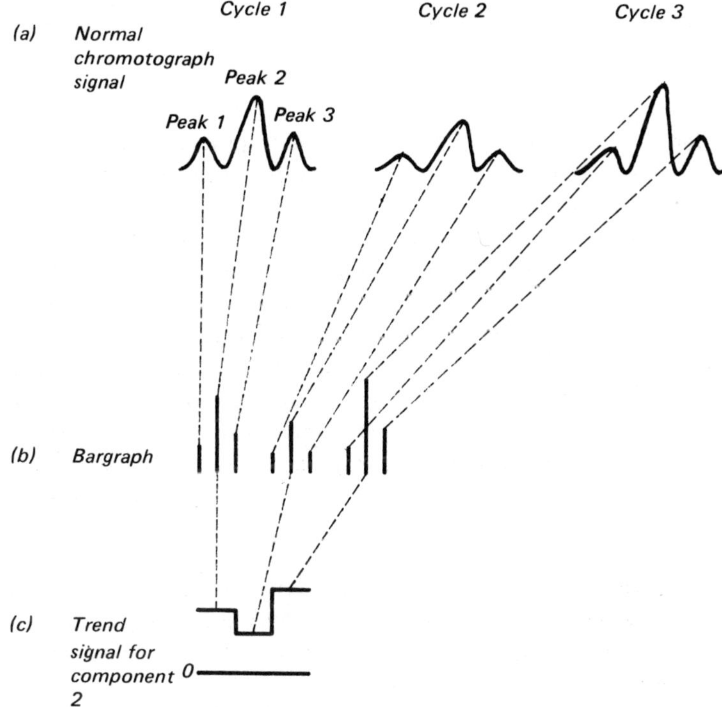

Fig. 10.1   Forms of signal presentation.

CHAPTER 11

The use of digital systems with process analysers

The most important applications of digital systems to process analysers are in data acquisition, conversion of the signal into useful engineering units, and presentation of results.

Digital equipment can also be used to program the operation of an analyser. This is particularly useful for application to chromatographs, because the timing of the gates and valves has to be adjusted carefully, relative to the elution times of the various peaks. These elution times vary with column age.

A differentiation must be made between 'built-up' or 'hard-wired' digital logic systems and the use of 'soft-wired' computer systems which have general-purpose logic, and can be adapted with special interfacing hardware and programmed with software to service analysers. Systems with limited calculation and presentation capabilities can now be built cheaply enough to service single analysers.

Before the present generation of highly reliable low-cost microprocessors was introduced, it was necessary that a computer system should service several analysers. In the case of chromatographs, to be economical, a system had to service at least seven chromatographs with peak-picking or three chromatographs with spectral-analysis.

Shared monitoring is normally acceptable, provided that it is reliable and accurate. However, for control, dedicated systems have long been preferred; thus advantage is now being taken of the availability of microprocessors.

Data may be presented digitally on an indicator or printed by a data logger. The display normally required is time, channel and value. The channel number may be a sequential number allocated to a particular stream and its components. Alternatively, it may be a tag number, a stream number, or a component or quality description with units.

Digital equipment is expensive so it must be used economically; only essential data should be presented. It is a common fault to include displays for components that have no importance, merely because they have significant concentrations. However, they may be useful for normalization purposes in some applications.

If the logger prints onto a prepared chart, the stream is identified by a location on the chart. It may only be necessary to print the time at the beginning of a line or, perhaps, even at the beginning of each new chart.

The units must be selected according to the final use to be made of the data. If a trend is to be observed, then it is sufficient to print the value of a concentration in mole % or volume % if the fluid sample has only one phase, but if the data are to be used for process-unit mass-balance, then density correction must be included in the conversion calculation so that a weight % print-out may be obtained. This assumes that most analysers give an output signal proportional to the molar concentration of the component. Some analysers, however, do give a weight % output signal directly. Analyser sensitivity will vary according to the stream and the component. Thus automatic correction must be included in the conversion calculation. Automatic calibration and sample handling can also be controlled by the computer.

As well as for data logging, the computer may be used for:

(i) averaging of signals over different periods of time according to the purpose;

(ii) calculations for the various balances, accounting and supervisory control;

(iii) the efficient reporting of analyser availability and utilization to improve maintenance and thus availability.

Operators want to check the actual volume flow-rate of a component whereas process engineers want to check the mass-balance. Nowadays either or both may be obtained, at the touch of a button.

A computer will normally read a signal from an analyser according to the availability of that signal. If the signal is continuously available, the analyser is read as a normal continuous sensor on a regular timed basis, e.g. every minute. More frequently the analyser gives a discontinuous signal. In this case either the computer must be informed by the analyser that a signal is ready to be read (a 'come read' signal is given) or the analyser must be read on a regular basis until the computer detects a new reading. The former case is more common and the diagram below shows how such a system works with a chromatograph. Similar systems are used with distillation, flash-point and many other analysers.

The period between computer-reading pulses can be selected according to the duration of the data impulse. If the amplitude of a 3-sec signal is to be read then the period should be 1–2 sec. If, however, a trend signal is available and if the period between the trend updating times is of the order of several minutes, the computer-reading pulse repetition period could be about 1 min with a consequent saving of computer capacity. The pulse width is usually about 10 msec with mercury-wetted relays, and several μsec with solid-state switching systems.

Smoothing by averaging the signal levels from a group of pulses should not normally be necessary, because efficient filtering should remove most of the noise and mains pick-up. Signal lines must be screened and the cables must be kept well away from power cables.

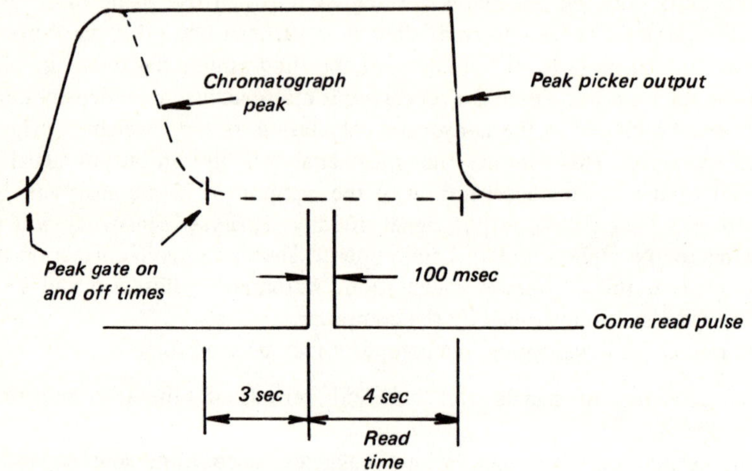

Fig. 11.1 – Beckman 6700 chromatograph programmer 'come read' pulse timing.

It is often necessary to earth an analyser signal at the analyser and it is always necessary to earth the input signal to the analogue-to-digital converter (ADC) at the ADC unit. Thus isolating units are often necessary.

Accuracy has been improved considerably by the use of better methods of averaging and noise reduction and more exact methods of linearization and compensation.

By the use of microcomputers and other modern electronic devices, it is possible to extract more useful information from data which were previously difficult to analyse.

Sometimes an analyser is used in conjunction with a dedicated computer (Fig. 11.2). The central processor unit (CPU) controls the operation of the analyser, e.g. programmes a temperature gradient or adjusts the signal attenuation in gas chromatography, and accepts and processes the results, performing computations such as reduction to standard conditions and normalization. There may also be a fault analyser, which normally only displays the faults of the computer system. Sometimes the actual fault has to be deduced from the display. The operating conditions, particularly the malfunctions, of the analyser system would normally have a separate display system. All faults preventing normal operation of the overall system should be logged together with the associated pre-alarms, and the time and date of the occurrence and repair of the fault.

Recorders are still needed for many systems since plant operators like to observe trends.

The CAMAC system has been developed for the interconnection of systems and digital computers. Many companies make this system, often designed for specific applications. As a result, certain program cards may be available only from a particular company.

The NIM system has been developed as a number of analogue cards using a system of packaging similar to the CAMAC system.

Modern analysers which utilize the latest digital techniques are becoming more difficult for the user to understand. Manufacturers should remember that operation and maintenance manuals must be very carefully written to ensure that they are adequately understood, as otherwise the analysers may fall into disrepair and bad repute. In order to improve documentation and thus maintenance, standard logic symbols should be used.

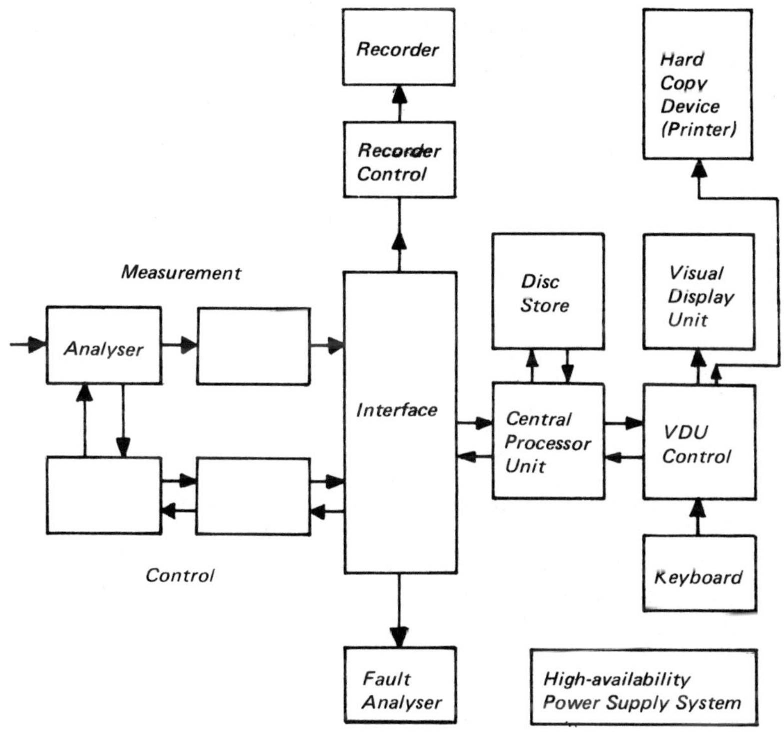

Fig. 11.2 — A typical system in which the analyser is operating with a dedicated computer.

CHAPTER 12

Control problems

Difficulties may arise if the components of a control loop are connected in the wrong order. For example, if peak-picked trend signals are used for resetting a controller with derivative-action controls, the trend-signal will be differentiated.

This undesirable differentiation can be prevented by injecting the cascading reset signal after the inner loop controller instead of before it (Fig. 12.1).

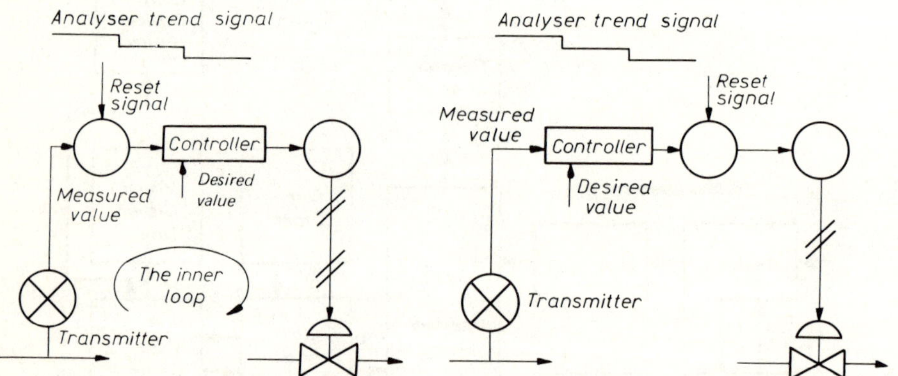

Fig. 12.1 – Problems and solutions when control loops are set by analyser trend signals.

The inner control loop must be faster than the outer resetting control loop. It is usually best to use an inner-loop controller with derivative action and an outer loop controller with integral action, as this is responsible for reset (Fig. 12.2).

The analyser control system of Fig. 12.3 illustrates the use of separate controllers for reset of a slave temperature controller and for direct proportional control of quality.

Sometimes the analyser is not considered to be sufficiently reliable for direct control because it may reset the operating point of the process plant beyond the required bounds, to a most inefficient or perhaps even dangerous condition. The analyser is often only required to reset the operating conditions of a process unit over very narrow limits to obtain improvement in plant efficiency. Thus, if the analyser signal goes beyond the predetermined limits, it could be arranged that the excess signal does not reach the controller.

The low-limit relay uses a high-value selector, which passes only the higher of the two input signals. Thus, the output signal cannot go lower than the low limit set by the low-limit regulating valve.

This is followed by a high-limit relay, which causes a low-value selector which passes only the lower of the two input signals (Figs. 12.4, 12.5). Thus, the output signal cannot go higher than the high limit set by the high-limit regulating value. Similar electronic devices are also available.

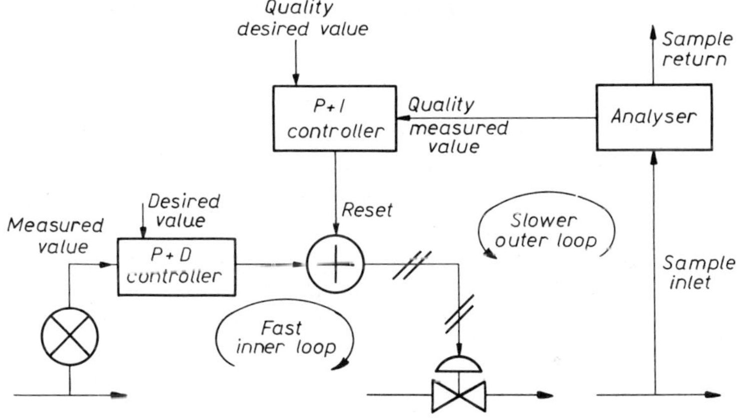

Fig. 12.2 — Analyser control of process variables.

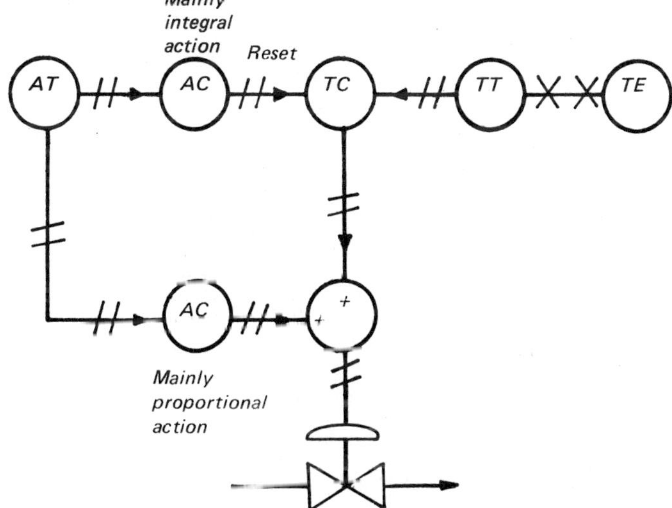

Fig. 12.3   Separate analyser controllers for resetting the temperature controller and direct regulation of the valve.

In a sampling control system, such as is used with a chromatograph, it is futile to try to regulate the process fluid more rapidly than the measuring system can respond. If the control is too fast, the system could go into an unstable lag oscillation.

If the chromatograph is measuring the overhead stream of a fractionator and is resetting the reflux ratio, and if the reflux ratio is large and the tower has a large number of plates, the process response lag — as well as the analysis lag — will be very long and it may take several days, or even weeks, to adjust the controller correctly.

To obtain good and fast control, the regulation should be made near to the sample take-off point. For instance, in a fractionating tower the overhead quality should be controlled by regulating the overhead conditions, such as pressure or reflux rate, and the bottoms quality should be controlled by regulating the bottoms conditions, such as temperature.

Unfortunately, variations at the top of the tower cause others at the bottom and vice versa. Therefore, the complete response can be quite prolonged and care has to be taken in the selection of sample points and regulator locations, if the complete response time really does matter.

Analyser controllers often require very long integral action times which are not easily attainable with normal components. A continuously rotating motor can act as an integrator. This is used in the Kent Boundless controller.

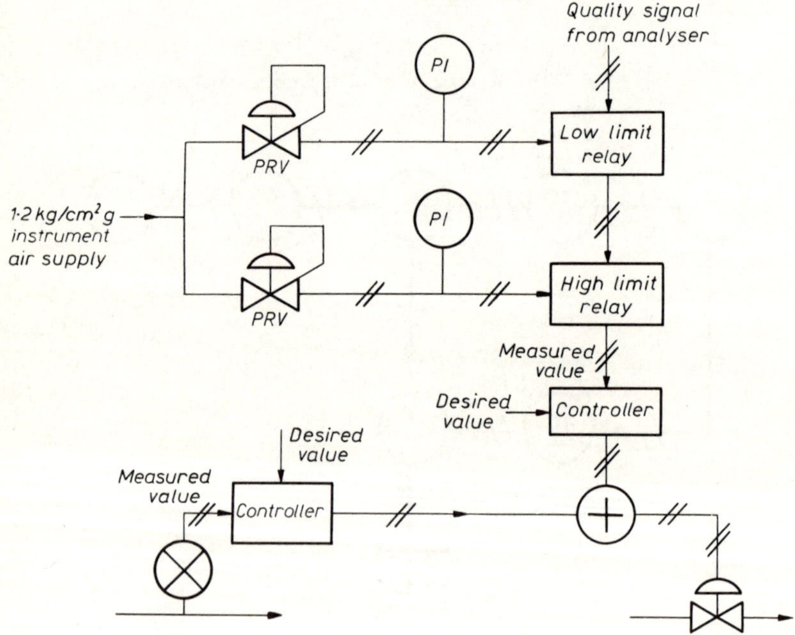

Fig. 12.4 – Use of relays to set analyser-signal limits.

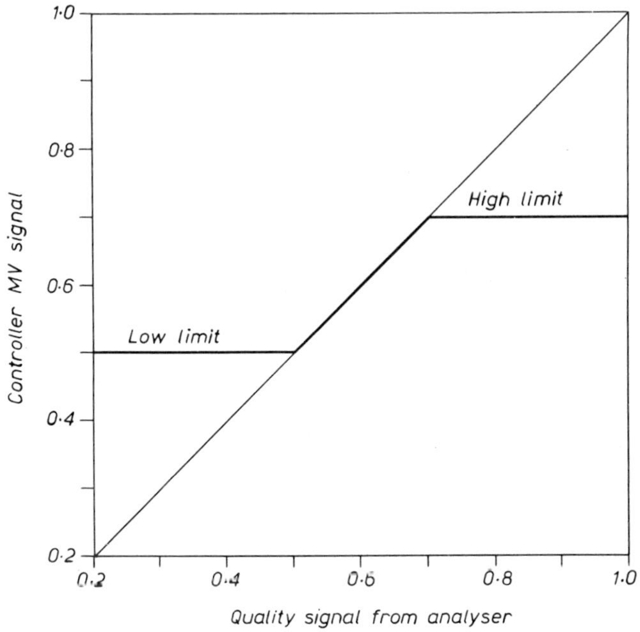

Fig. 12.5 — Signal limits.

The analyser measures the concentration of a component. The quantity of the component is the concentration of the component times the rate of flow of the fluid of a particular stream. Thus, to control the addition of a reagent in proportion to the quantity of the components in the stream, the system shown in Fig. 12.6 is required. The system has been applied to control the quality of a fractionator-column overhead by regulating the reflux ratio.

In the system shown in Fig. 12.7, the analyser quality deviation signal, from the quality controller, resets the reflux ratio.

In the system shown in Fig. 12.8, the same quality-signal should be obtained as from the previous system, provided that there are no uncondensed gases. The liquid sample has been returned to the process unit because of its appreciable mass flow-rate and resultant commercial value. Use is made of the pump pressure head; sufficient pump capacity above the process requirements is usually economically available.

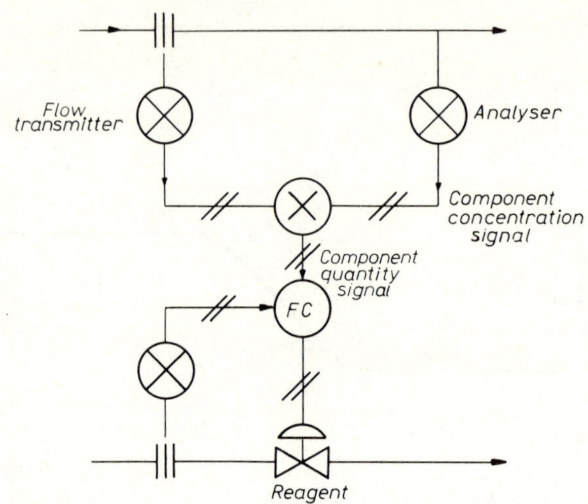

Fig. 12.6 — Reagent flow-rate set by quantity of component.

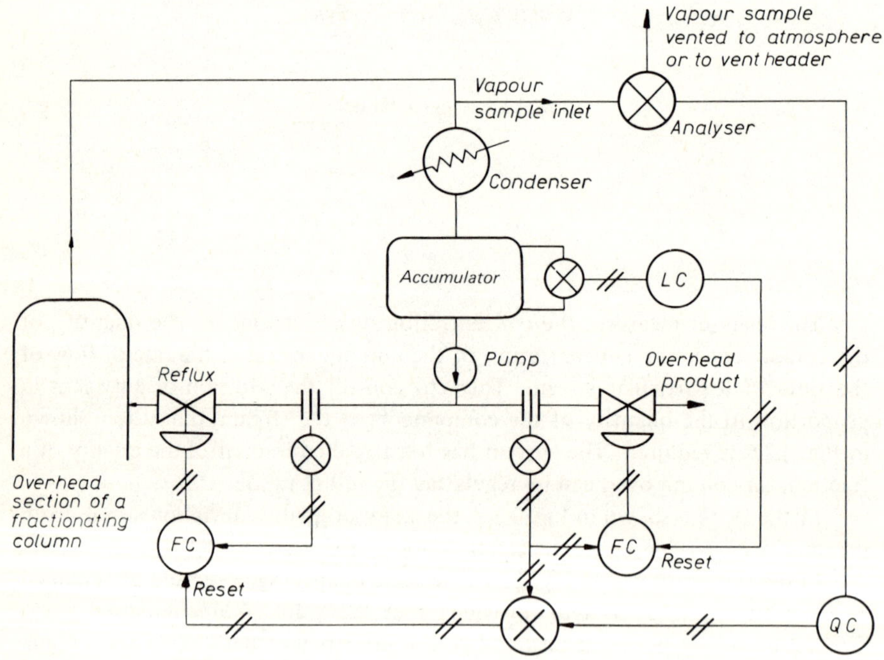

Fig. 12.7 – Controls for fractionation-column overhead.

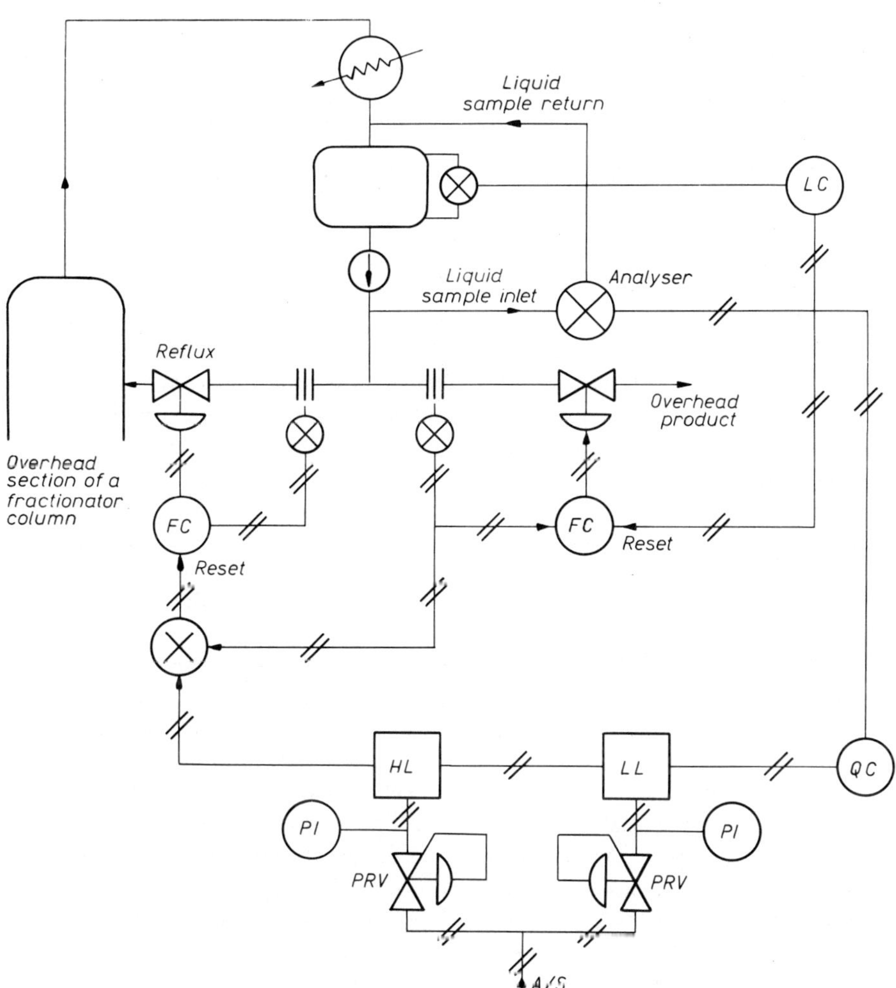

Fig. 12.8 – The use of signal limiting (HL = high level limiter, LL = low level limiter).

CHAPTER 13

Maintenance and availability

An analyser is said to be *available* when it produces an acceptably repeatable output signal.

The *availability* of analysers with adequate specialist maintenance is of the order of 80–90%, depending on the quality of the maintenance, the analysers and the sampling systems, and the cleanliness of the samples. With general instrument maintenance, the availability is typically only about half that obtainable with specialist maintenance.

Each specialist maintenance technician should be capable of looking after 10–20 analysers, depending on the complexity, reliability, and suitability of the analysers for the particular applications. A badly-designed sample system handling a dirty sample will require much more attention than a well-designed sample system handling a clean sample.

The initial availability of an analyser depends on the training of the maintenance technician and on the running period of the equipment in the workshop. Practical training on individual analysers should start with factory visits during the works tests.

Familiarity of the technician with the analyser increases with time, but performance of the analyser decreases with time. As a result, availability first increases with time as familiarity improves; debugging also improves availability. The availability reaches a maximum which is dependent on the number of analysers maintained by each technician. Fifteen analysers per technician results in a peak availability of the order of 80%, whereas a reduction to 10 analysers per technician results in a peak availability of the order of 90%. Higher availabilities of the order of 95–98% have been reported for new analysers on new plants where maintenance is well organized.

The rate of decrease of the availability over a long period of time can often be reduced by periodic visits by the analyser manufacturer.

Both the instantaneous and the average availability need to be known.

The instantaneous availability is that at the time of reporting: the analyser is said to be instantaneously available or not. The average availability is determined from the time of installation and does not mean very much until a period of at least 6 months has elapsed.

Analyser utilization should also be reported. An analyser may be available but not in use, perhaps because a particular plant is down or because a certain product is not being made or delivered.

The mean time between failures, for most analysers, varies between 6 and 12 months, depending upon the type of analyser and the quality of the maintenance and of the sample stream. An analyser with many rapidly moving parts will have a shorter life than one with no moving parts.

The down-time of an analyser when it is being maintained is very short nowadays. It is of the order of 1–4 hr and can be minimized if all the necessary facilities are available in the analyser house. The down-time can be much longer if only a shelter is used and the analyser has to be disconnected and taken to the workshop for maintenance.

Many analysers take many hours to be set up properly, depending on their complexity and on the number of components and streams being handled.

Analysers should not be connected to the process stream immediately a plant comes on production but should be left out of service for a period, to allow pollution to clear from lines and vessels. This also applies to new plants on first commissioning, when the analysers should always be commissioned last of all. Many complications, damage and renewals of components will be avoided if this procedure is followed.

The time can always be fully and profitably utilized by training, checking and testing with prepared samples, wherever this is practicable.

Sometimes a successful plant start-up depends on the analyser being available. The sample system should then be designed with this in mind.

The *usefulness* of an analyser system is defined with respect to the increase in the efficiency of a process unit resulting from the incorporation of the analyser system. The *effectiveness* of the analyser system is defined as the relative increase in the efficiency of the unit.

Let us now try to relate availability (% up-time), mean time between failures (MTBF) and mean time for repair (down-time) for a typical example.

Assume that the mean time between failures is 6 months and that all spares are held in stores or are easily available from the vendor. Also assume that it takes one week to detect a fault in an analyser and to get it into the workshop, and that it takes another week to repair the analyser and return it to the process unit:

$$\text{Availability} = \frac{\text{MTBF}}{\text{MTBF} + \text{down-time}} \times 100\%$$

$$= \frac{26}{26 + 2} \times 100\% \simeq 93\%$$

If it takes three weeks to get the new component because it was not immediately obtainable from stores then

$$\text{Availability} = \frac{26}{31} \times 100\% \simeq 84\%$$

In practice, high availability can be maintained with reliable analysers working in a suitable atmosphere with dry and clean air, clean cool water and stable noise-free electrical supplies, provided that the filters and strainers are regularly cleaned or the elements replaced, that all orifices are kept clear, and that coalescers are properly adjusted. Provided that regular and thorough maintenance visits are paid to each analyser, there is no reason why availability should not be as high as 95% or even higher.

For normal instrumentation, preventive maintenance is usually found to be uneconomical, so breakdown maintenance is more normal. However, for some analysers, especially those on critical control and alarm duties, preventive maintenance is essential so the maintenance demand must be carefully recorded, controlled and minimized. The total maintenance demand is equal to the sum of the breakdown maintenance demand in hours and the preventive maintenance demand (hours per visit × number of visits per year). Sometimes it is informative to break down the demand into the components inspection, overhaul, repair, calibration etc.

The description of the failed component and the date of failure should be noted on the analyser maintenance record card.

If analyser-maintenance mechanics are not available, it is worth while having a maintenance contract with the analyser vendor or with a suitable specialist if one is available. When the analyser and plant have been commissioned, the contract should be for monthly visits. Between these visits, if the equipment fails it should be left until the contractor is available. If untrained mechanics try to repair the analyser, they can easily cause more damage.

If the mean time between failures is proved to be other than a month, then the period between visits should be adjusted accordingly.

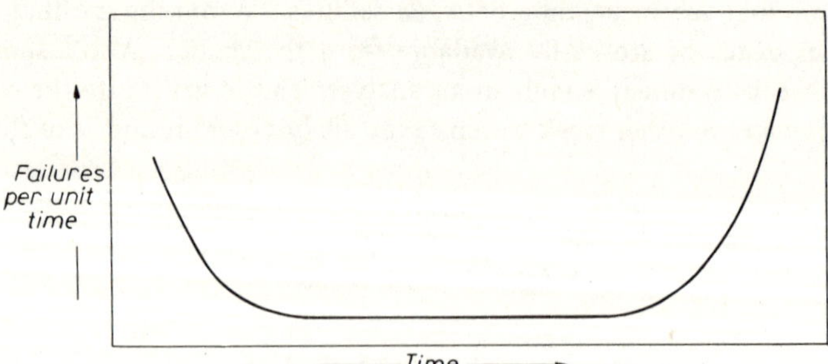

Fig. 13.1 — The variation of failure during lifetime.

A useful rule is, 'Make one person responsible for the equipment and thus reduce knob twiddling'. Availability of the equipment then increases considerably. If the analyser-maintenance mechanic takes a really intelligent interest in the analyser, the availability of the analyser is high. If the users (operators) have no faith in the analyser, then sometimes it just is not worth while continuing to use it. The amount of analyser equipment in the control room should be minimized so that the users are not aware of the frequency of maintenance visits necessary even with good equipment.

The life of components or of equipment can be divided into three periods (Fig. 13.1).

  (i) The early failure or infant mortality period.

  (ii) The random failure period, when the failure rate drops to a low constant level. The fraction surviving is given by $e^{-t/m}$ where m is the mean time to failure.

  (iii) The wear-out period. During this third period the failure rate rises rapidly.

Reliability, availability and maintainability are defined below:

$$\text{Reliability} = R(t) = e^{-\lambda t}$$

where λ = failure rate = 1/MTBF, t = time duration over which reliability is required and MTBF = the mean time between failures.

$$\text{Availability} = A = \frac{\text{MTBF}}{\text{MTBF} + \text{MDT}}$$

where   MDT = the mean down-time

                = repair time + administration time + delay time.

$$\text{Reliability} = R(t) = e^{-t/\text{MTBF}}$$

$$\text{Maintainability} = M(t) = 1 - e^{-t_r/\text{MTTR}}$$

where t_r = repair time and MTTR = mean time to repair.

Repair time should be subdivided into the time for corrective repair and preventive maintenance, and the time that the analyser is out of service for overhaul.

Availability can be improved if the time spent during the various activities shown in Fig. 13.2 is analysed. Preventive maintenance and overhaul should be planned for minimum down-time.

In any attempt to improve system reliability by the use of more than one analyser to measure the same process quality (Fig. 13.3), it is most important that the failure of an analyser should be rapidly recognized and corrected. In addition to the more usual causes of failure which are annunciated, it would also be desirable to annunciate an out-of-limits deviation of the output signals.

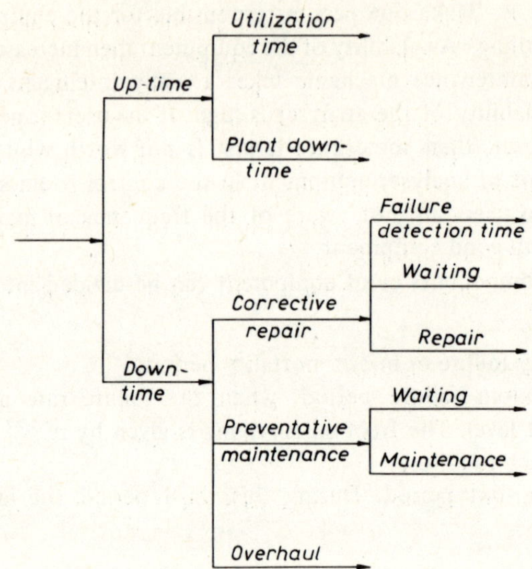

Fig. 13.2 – Analysis of time spent on various activities.

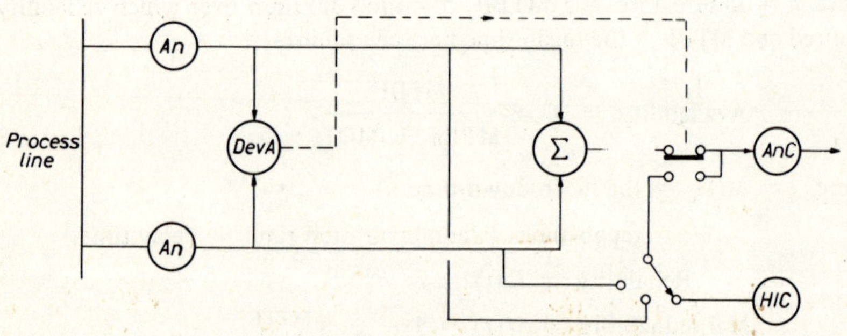

Fig. 13.3 – A typical redundant analyser system.

If the analysers are resetting a controller, it would be necessary for the deviation detector to switch the reset signal from remote to local manual. The operator could then switch the secondary controller manually to the signal from an analyser which is not faulty.

It is often simple to make the analyser-signal summator sufficiently reliable for it not to impair the overall system reliability.

The probability that a catastrophe will occur is equal to the product of the probability that a critical component is at a critical concentration and the probability that the system-protecting analyser does not detect the critical concentration.

For example, for an explosion to occur, the flammable component must be present at a sufficient concentration, sufficient oxygen must be present and sufficient electrical energy or heat must be present to ignite the flammable mixture. The probability of the occurrence of an explosion is given by

$$P_C = P_F \times P_O \times P_E$$

The classification of a hazardous area is dependent upon the probability of the occurrence of a critical concentration of a flammable gas in air.

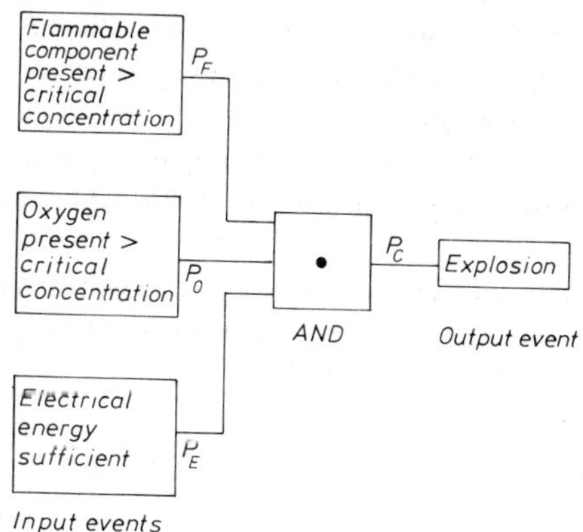

Fig. 13.4 – Probability of occurrence of a catastrophe.

Electrical equipment is then selected so that the product $P_F \times P_O \times P_E$ is less than once in several hundred years. In this case, it is assumed that air is always present, i.e. $P_O = 1$.

For a Zone 1 area, P_F is large ($> 10^{-4}$) so P_E has to be very small. For a Zone 2 area P_F is smaller (between 10^{-4} and 10^{-6}) so P_E can be larger than for the Zone 1 case. Here a probability of 10^{-4} means that the hazard exists for 1 hr in every 10^4 (i.e. for about 1 hr per year).

The probability of an occurrence can be reduced by common-sense precautions. For example, when an inert gas is used to prevent air coming into contact with a warm liquid, an oxygen analyser is often used to detect the onset of an excessive concentration of oxygen The probability of the occurrence of a dangerous concentration of oxygen is minimized by the use of a reliable and fast pre-alarm and shut-down system. The shut-down frequency must be very low, so the inert-gas system must be extremely reliable.

Again, when a burner has been extinguished, it should not be re-ignited until the flammable gases have been purged from the furnace. This is normally done on a time basis without an analyser.

The probability that an analyser system will fail is equal to the sum of the failure probabilities of the individual components in series and the product of the failure probabilities of the parallel components or systems, provided that the probabilities are mutually independent.

When there are several concomitant probabilities, they can be summed as illustrated in de Heer's paper.

The protection-loop power supply for such systems is often a battery-driven inverter. The solenoid valve should be de-energized on power failure, so that the air is then shut off from the valve.

The probability that an undetected or open-mode failure of a protecting system occurs at the same time as a hazardous condition in the plant must be extremely small, say once in more than 500 years, otherwise the resultant total number of calamities in the plant may be too high. On the other hand, a complex fail-safe system will probably have too many short-mode failures and will thus shut the plant down too often.

The protecting system must thus be as simple as possible and be inspected sufficiently frequently for the rate of open-mode failures to be acceptably small.

Example: If the open-mode failure rate is once per 100 years, and the plant hazardous condition can occur 4 times per year, then if inspection is every month and the mean repair time is 2 weeks,

$$\text{the MTBF}_o = 100 \times \tfrac{1}{4} \times \frac{52}{2} = 650 \text{ years.}$$

which should be acceptable.

If the short-mode failure rate is 4 per year,

$$\text{then MTBF}_s = \tfrac{1}{4} \times \frac{52}{2} = 6\tfrac{1}{2} \text{ years,}$$

which is probably too frequent.

BIBLIOGRAPHY

Browning, R. L., Loss analysis improves process protective instrumentation, *Instrum. Technology*, 1972, **19**, No. 10, 27 October 1972, pp. 27–32.

de Heer, H. J., Calculating how much safety is enough, *Chem. Eng.*, 1973, 19 February, 1973, pp. 121–128.

Cason, R. L., Estimate the downtime your improvements will save, *Hydrocarbon Processing*, January 1972, pp. 72–76.

CHAPTER 14

Examples of applications of analysers

14.1 INTRODUCTION

In order to select the component to be determined for controlling a reactor it is necessary to examine the concentration of the reaction components, the concentration of the interfering components, and the condition of the sample. An example of a reactor control system is given in Fig. 14.1.

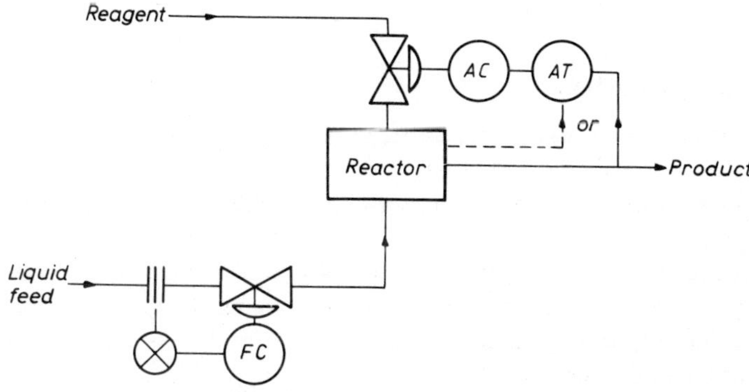

Fig. 14.1 – The control of a reactor.

If the feed is a liquid and the reagent is a gas, the sample may be taken from the vapour space and the unreacted gas concentration measured. This concentration may be rather small if the gas is soluble in the liquid. Alternatively, the concentration of the unreacted gas in solution may be measured, but there may be interference from the product. Product purity measurement may thus be an acceptable alternative. Secondary reactions may occur, however, which produce interference with the measurement, or a precipitate may be formed, etc.

If the reaction takes time to complete, it is necessary to locate a sample take-off point where the reaction is sufficiently complete and the lag time is acceptable for control.

Some analysers are required to monitor the impurities in a product stream after fractionation or after a reaction. Other analysers may be required to analyse the same stream in order to control the fractionator or reactor so that impurity concentrations are minimized as quickly as possible, consistent with operating efficiency. In this last application, the fractionation or reaction need not be complete. Two separate analysers may not be justified for the two streams, and a two-stream analyser might be acceptable if the analyses are sufficiently fast for the control application.

Examples may be found in the following production-stream measurements and loop controls.

(i) Fractionator overheads with the analyser resetting the reflux flow-rate or ratio.

(ii) Fractionator bottoms with the analyser resetting the temperature of the reboiled vapour or of the liquid near the bottom of the column.

(iii) The controlled chlorination of dirty water, with acceptably low residual chlorine in the effluent.

(iv) Controlled neutralization to satisfy requirement for a neutral effluent.

(v) Detoxification of effluents from plating works, etc.

14.2 FRACTIONATION

In liquefied petroleum gas (LPG) units, process gas chromatographs are used to monitor the overheads and bottoms products and to control the fractionators for optimum economic operation with acceptable product impurity levels. (Figs. 14.2–14.4).

Multi-stream chromatographs are usually not preferred, because of the possibility of impurity interference between streams. It was for these applications that the Elliott LPG analyser was designed by Topham and others at the B.P. Isle of Grain refinery.

The bottoms sample should be taken from a location after the cooler to ensure that the liquid sample contains no vapour phase. The off-gas sample line often requires to be heat-traced to prevent condensation.

If the concentrations of the impurities on each side of the major components are measured, one chromatograph could be adequate for control of a fractionator, provided that the feed analysis is sufficiently constant.

Sometimes the required purity of a stream from the extremity of a fractionation tower is so near to 100% that the concentration of the impurities cannot be measured sufficiently precisely. In this case the precision can be improved by analysing the fluid nearer to the feed nozzle where the concentration of the impurities is greater (Fig. 14.5).

Figure 14.6 illustrates the likely location of some analysers in the main fractionation towers of a refinery.

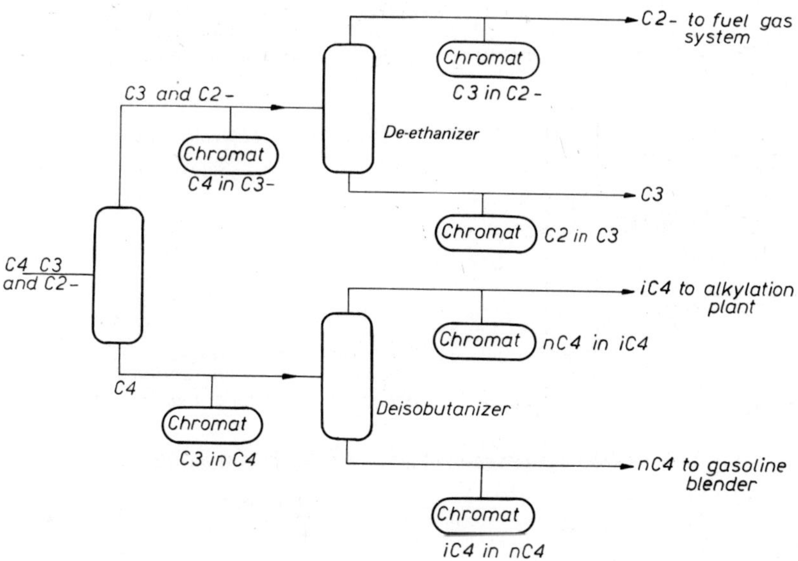

Fig. 14.2 — A typical LPG fractionation plant.

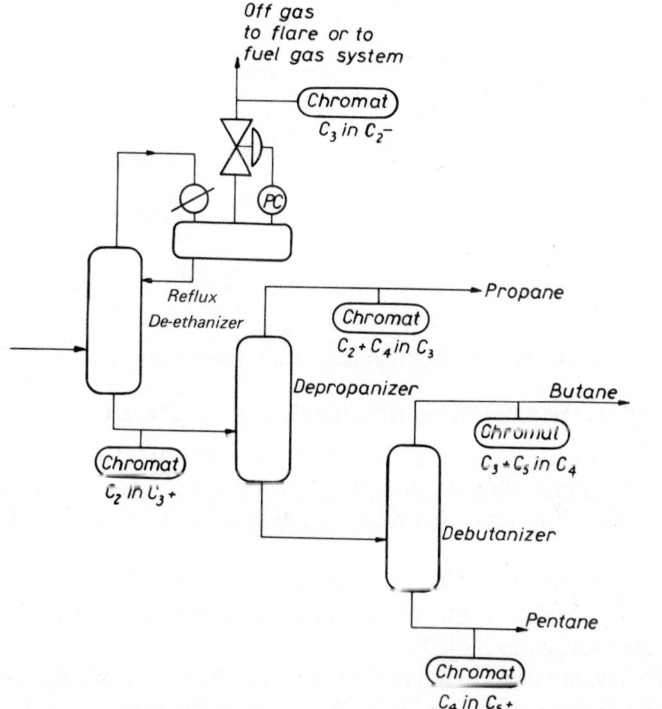

Fig. 14.3 — Locations of chromatographs in a typical LPG fractionation plant.

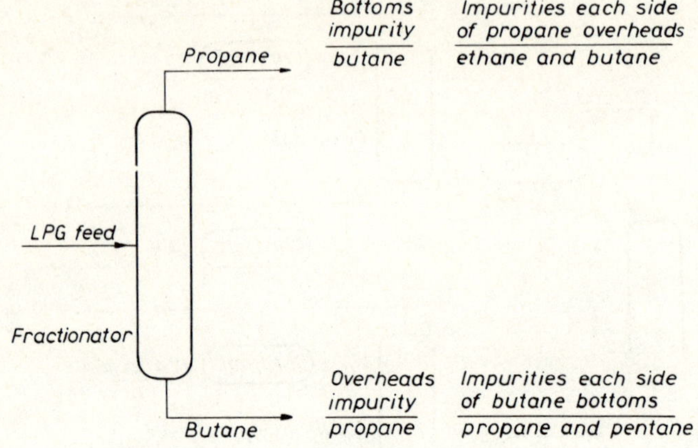

Fig. 14.4 – Components measured in overheads and bottoms streams.

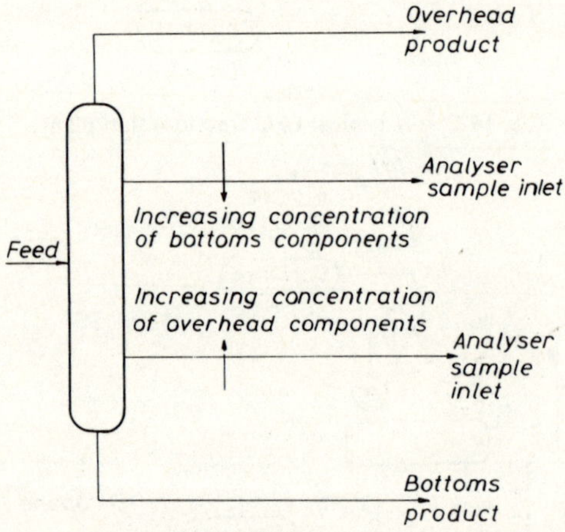

Fig. 14.5 – Analysis of fluid near to the feed nozzle.

14.3 BREWING OF BEER AND DISTILLING OF ALCOHOL

In the brewing of beer, it is usual to measure the density of the wort or of the beer within the range 1000–1100 kg/m^3, which corresponds to a range of 0–100 brewer's degrees. The result is usually required to be accurate to within 0.03–0.1%.

Section 172(2) of the U.K. Customs and Excise Act 1972 required that the Original Gravity (OG) of the wort should be measured as $SG^{60°F}_{60°F}$; this has recently been changed to $SG^{20°C}_{20°C}$.

Density measurement is also used during and after the distillation of alcohol, as a measure of the alcohol content of the product. The main impurity is usually water. Table 14.1 gives the relation between alcohol content and specific gravity.

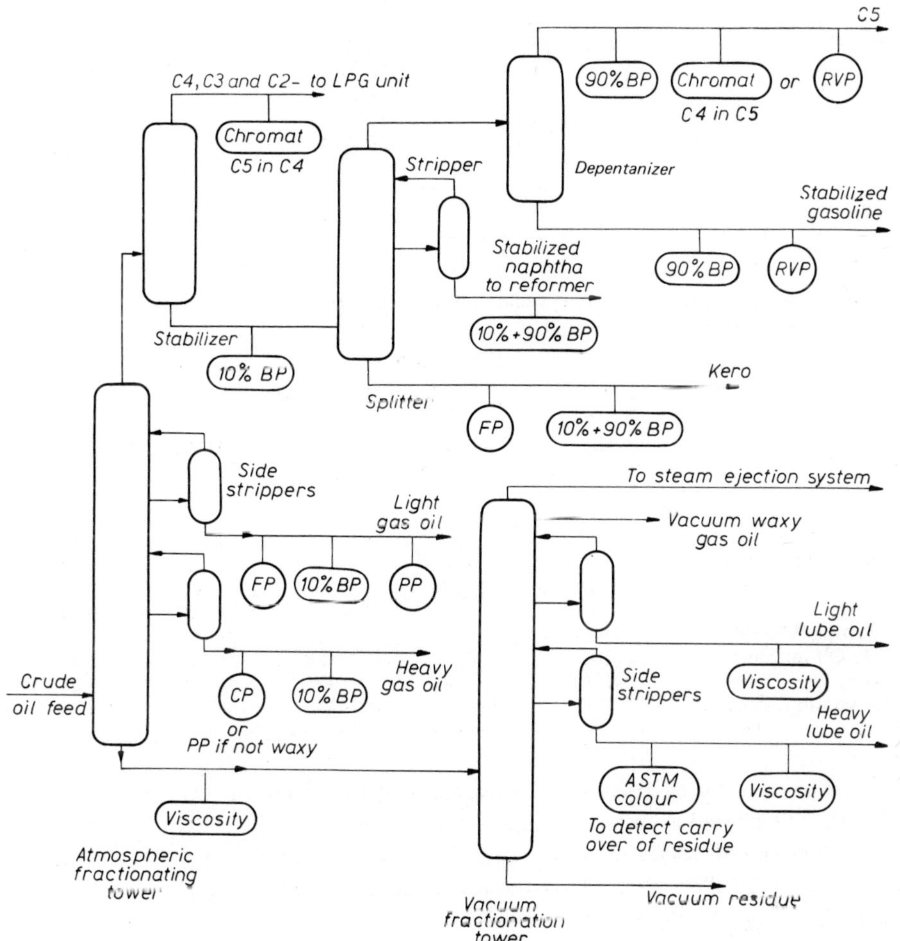

Fig. 14.6 – Location of analysers in the main fractionation towere of a refinery (BP = boiling point, RVP = Reid vapour pressure, PP = pour point, CP = cloud point).

Table 14.1

Specific gravity of aqueous ethanol solutions

Ethyl alcohol %w/w	$SG^{20°C}_{20°C}$
100	0.79040
98	0.79658
96	0.80250
94	0.80825
92	0.81383
90	0.81925
85	0.83228
80	0.84483
75	0.85702

Table 14.2 gives some typical values for the specific gravity of various alcoholic products.

Table 14.2

Specific gravity of alcoholic products

Liquid	U.S. Proof Spirit Scale	U.K. Proof Spirit Scale	$SG^{60°F}_{60°F}$	Alcohol Content	
				% v/v	% w/w
Beer (1034.5 Gravity)	–	5.48	0.9954	3.13	2.51
Beer (1055 Gravity)	–	9.03	0.9926	5.18	4.14
Very Strong Beer	–	15.38	0.9880	8.80	7.08
	50	–	–	25.00	20.44
Typical Whisky	–	70	0.9518	40.00	33.37
U.S. Proof Spirit	100	–	0.93428	50.00	42.49
U.K. Proof Spirit	150	100.00	0.91976	57.10	49.28
Absolute Alcohol	200	175.35	0.79360	100.00	100.00

For an error of 0.1% in the measurement of density or S.G. the temperature of the liquid has to be maintained constant to better than 1°C.

U.K. proof spirit at 51°F (10.5°C) weighs 12/13 as much as an equal volume of distilled water. It is the lowest concentration of alcohol which will allow ignition of gunpowder moistened with it.

14.4 DEHYDRATION

The life of a drying agent may be extended considerably by drying the product only to the required moisture content. This may be achieved by controlling the rate of flow of the fluid which by-passes the drying agent. The desiccant must be regenerated when the moisture content of the product has reached the preset alarm level.

The type of moisture analyser and the details of the sample conditioning system have to be selected carefully according to the process fluid and the type of drying agent. Different types of analyser are usually used for gas and for liquid streams.

Acid gas containing hydrogen sulphide reacts with phosphorus pentoxide, as also does caustic soda which may be present in the water to be removed after the treatment of the process fluid to remove residual sulphur compounds.

It is very difficult to maintain an acceptably low glycol carry-over from a glycol-type drier. A typical carry-over is 10–30 ng/ml; even this low concentration is too much for some systems.

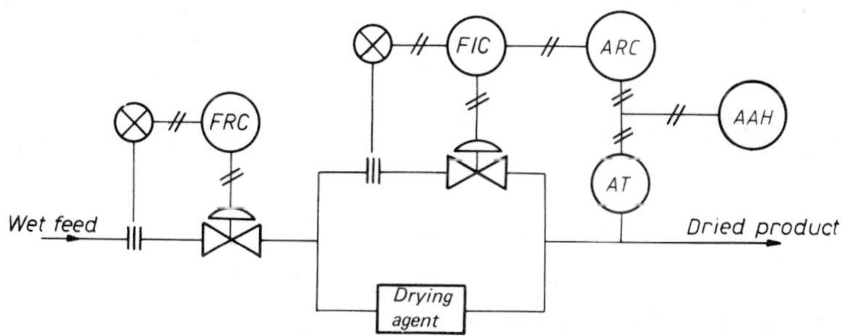

Fig. 14.7 – Moisture control.

14.5 COMBUSTION

The maximum efficiency of combustion of fuels in furnaces is reached by adjusting the fuel–air ratio so that there is a minimum excess of air with only a very small amount of smoke coming from the stack. This is obtained by adjusting the air–fuel ratio from the oxygen-analyser signal so that there is a pre-determined excess of air.

It is necessary to take a truly representative sample from the stack before any air has been sucked into it (since it would interfere with the measurement).

Ideally, the gas in the vicinity of each flame should be analysed, so that the air–fuel ratio for each individual burner or group of burners is separately adjusted.

For combustion control of furnaces and boilers, the measurement of the oxygen concentration in the flue gas is preferred to carbon dioxide measurement.

The CO/CO_2 ratio in the flue gas is less dependent on admitted air than is oxygen concentration.

The oxygen in the flue gas is a better measure than carbon dioxide for the excess of air because it is less dependent on the C/H ratio of the fuel (Fig. 14.8).

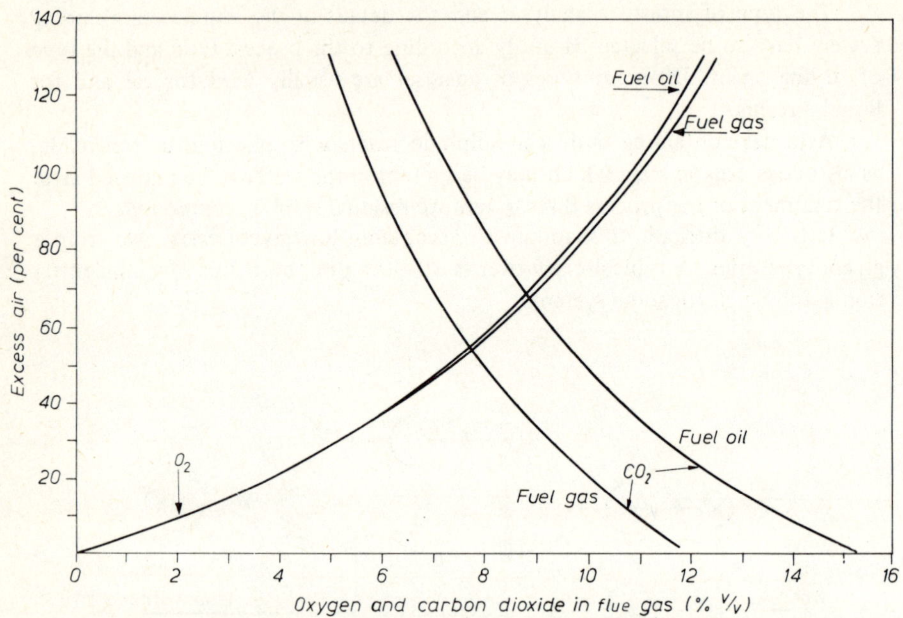

Fig. 14.8 – Oxygen and carbon dioxide in flue gas.

14.6 CRUDE-OIL STABILIZATION

Crude oil is stabilized by the removal of LPG, which often contains a large amount of H_2S and other sulphur compounds, most of which must be removed to protect the LPG plant and to produce saleable propane and butane (Fig. 14.9).

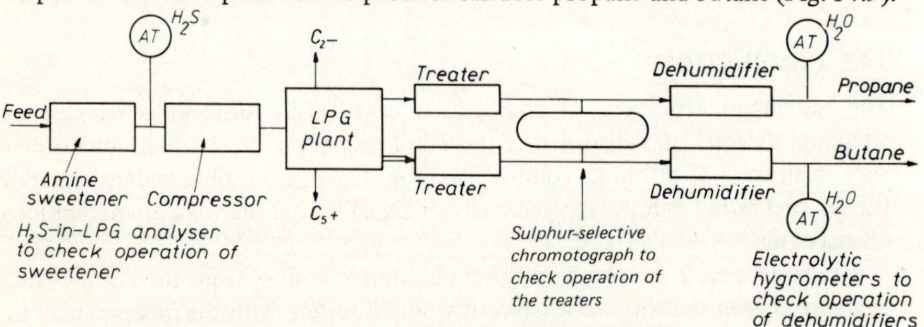

Fig. 14.9 – Control of removal of sulphur compounds. The chromatograph has a sulphur-selective flame photometric detector, range 0–25 μg/g.

Most of the H_2S is present in the de-ethanizer off-gas unless amine sweetening is required upstream of the de-ethanizer. Further removal of the remaining sulphur compounds from the refined propane and butane is usually necessary. A Reid vapour-pressure analyser is installed to ensure that the crude oil has been sufficiently stabilized. For accounting purposes, the density of the crude oil is required to be known very accurately.

14.7 NAPHTHA REFORMING

In the recycle gas stream of a catalytic hydrogenation reactor; the hydrogen concentration is a measure of the amount of hydrogenation occurring and thus of the amount of refined hydrocarbon in the product stream. The hydrogen concentration can be measured by a selective thermal-conductivity analyser or by a gas-density analyser, which can also be used to correct the normal recycle gas-flow measurement for gas-density variations. The Foxboro gas-density cell with its associated differential-pressure transmitter has been used successfully for this purpose, with established long-term reliability.

A simplified block-diagram of a reformer is shown in Fig. 14.10. Up to 1970, it was not common to use an octane-rating analyser to set the reactor-feed temperature. Manual resetting of the temperature controller was usually sufficient. With modern increased severity of reformation and the availability of the UOP and Anacon continuous analysers, closed-loop analyser control should become more common.

With a new catalyst, a naphtha reformer works at about 500°C and produces about 70% of hydrogen in the recycle gas. As the catalyst ages, it becomes coked up, the octane rating of the reformate is reduced and the hydrogen concentration in the recycle gas is also reduced. To overcome this, the reforming temperature is increased to a maximum temperature of 525°C. The higher the temperature, the more rapidly is the carbon laid down on the catalyst.

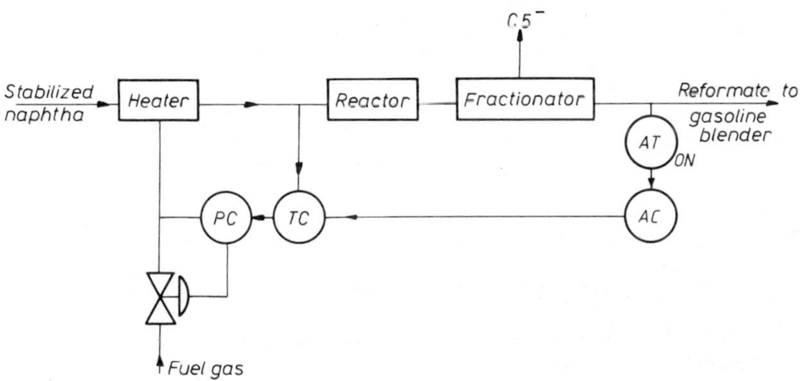

Fig. 14.10 – Reformer unit control based on reformate product octane rating setting the temperature of the reactor feed (ON = octane number).

14.8 CATALYST REGENERATION

Reformer catalyst usually has to be regenerated when the hydrogen concentration in the recycle gas has dropped to about 55% (Fig. 14.11). To do this the carbon is oxidized to carbon dioxide by passage of air through the reactor. To start the reaction, the air has to be preheated to a certain temperature, above which the reaction is exothermic, but care has to be taken to prevent excessive temperatures occurring.

One method uses restricted air flow controlled by the measurement of the oxygen concentration in the outlet gas (Fig. 14.12) which typically increases 0.5 to 2.5% v/v during the regeneration process.

Another method uses a very large air flow to remove the heat generated, which is recovered in a waste-heat boiler. The concentration of the carbon dioxide in the exhaust gas decays rapidly from somewhat less than 20% v/v.

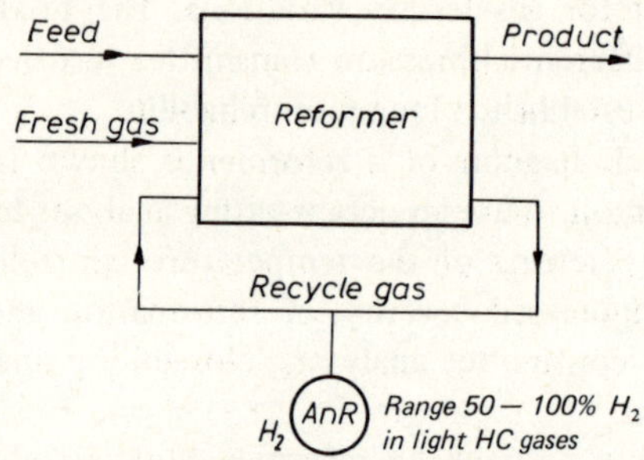

Fig. 14.11 – Measurement of the hydrogen content of reformer recycle gas.

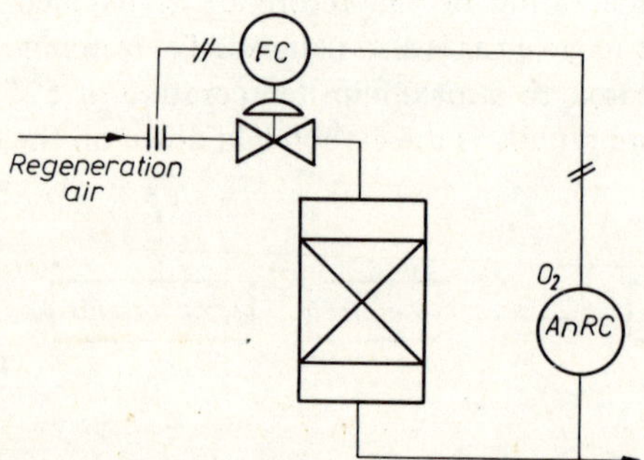

Fig. 14.12 – Control of catalyst regeneration by measurement of oxygen excess.

14.9 HYDROGENATION OF ACETYLENES

The hydrogenation of acetylenes to form paraffins may be controlled from an analyser signal (Fig. 14.13).

$$2\,H_2 + C_2H_2 \longrightarrow C_2H_6$$

$$2\,H_2 + CH_3C{\equiv}CH + CH_2{=}C{=}CH_2 \longrightarrow C_3H_8$$

A signal proportional to the concentration of acetylenes in the reactor outlet may be used to adjust the operating conditions of the reactor and to determine when the catalyst should be regenerated. A process gas chromatograph can be used for the measurement of this concentration, which is typically of the order of a few ppm.

The hydrogenation of acetylene can be controlled so that the product is ethylene, not ethane (Fig. 14.14). In this system, the hydrogenation of the feed stream is optimized so that the concentration of acetylene in the purified ethylene is kept as low as necessary without excessive hydrogenation of ethylene to ethane (which would represent product loss).

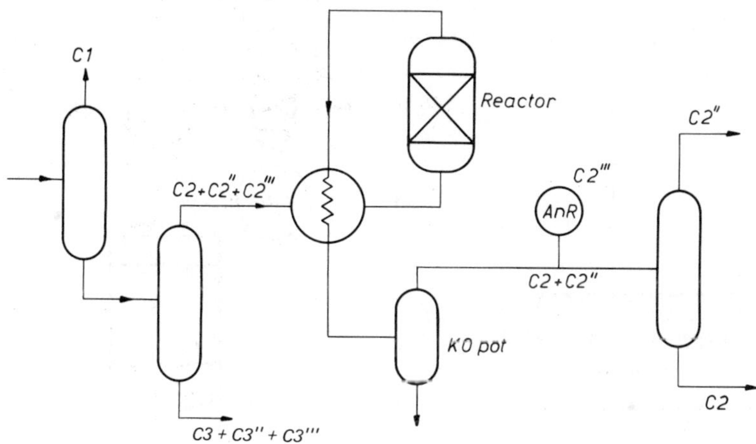

Fig. 14.13 – Acetylene hydrogenation unit (C = alkane; C" = alkene; C'" = alkyne).

14.10 DESULPHURIZATION

The proportion of the feedstock which has to be treated in a desulphurizing unit depends on the concentration of sulphur in the feed and the desired level of sulphur concentration in the product. Thus the flow-ratio controller can be set by the sulphur analyser so that the unit is most economically used (Fig. 14.15).

Sulphur is removed from hydrocarbons by catalytic hydrogenation of the sulphur compounds:

$$R{-}SH + H_2 \longrightarrow H_2S + R{-}H$$

This is followed by selective absorption of the hydrogen sulphide by an amine such as diethanolamine and then extraction of the hydrogen sulphide from the amine. The hydrogen sulphide is then passed to a sulphur recovery plant which is usually based on the Claus process. The concentration of the hydrocarbons in the feed to the Claus plant must be minimized in order to limit carbon build up on the catalyst.

In the furnace, approximately 43% of the hydrogen sulphide is oxidized to sulphur, and the rest to sulphur dioxide:

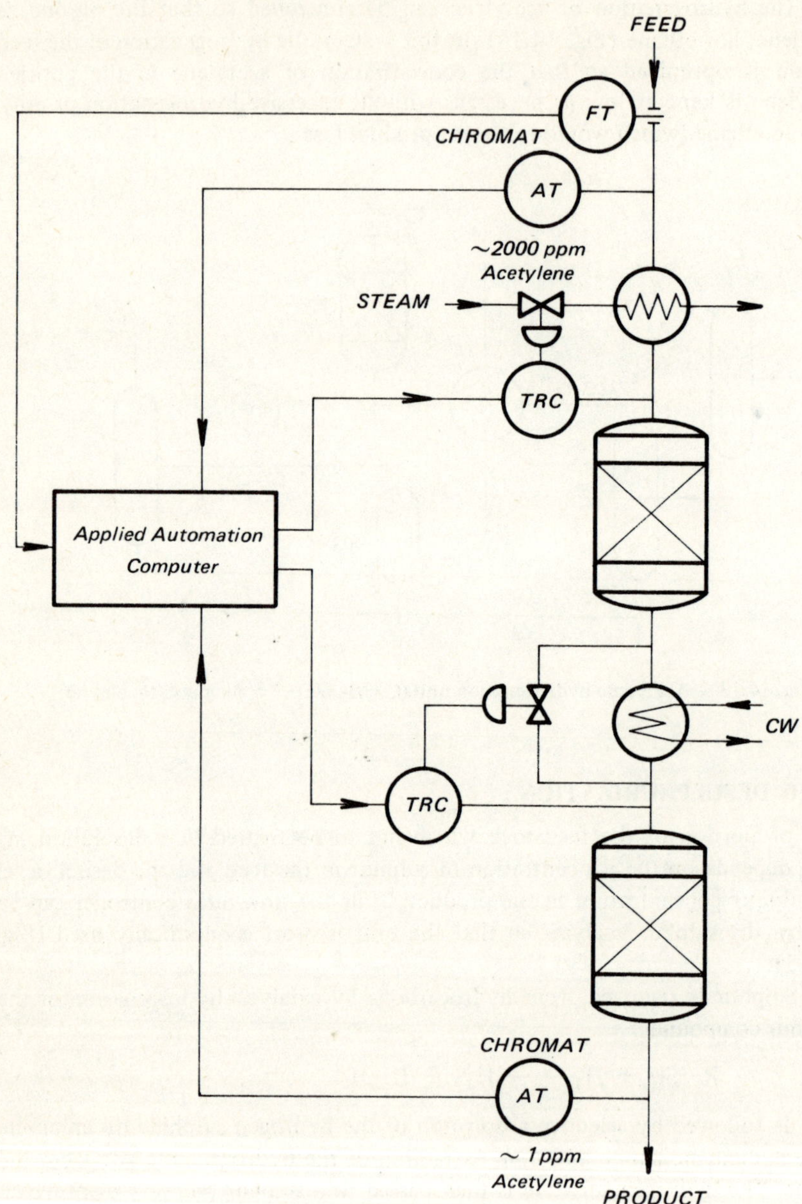

Fig. 14.14 – Acetylene hydrogenation reactor.

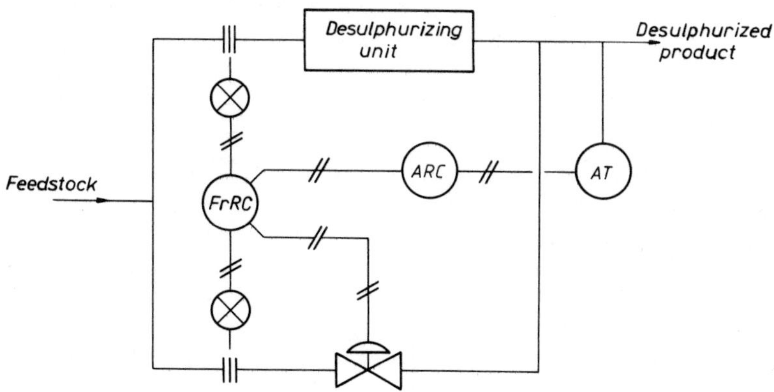

Fig. 14.15 – The most economic use of a desulphurizing unit.

$$H_2S + \tfrac{1}{2}O_2 \longrightarrow S + H_2O$$

$$H_2S + 1\tfrac{1}{2}O_2 \longrightarrow SO'_2 + H_2O$$

In the reactor additional hydrogen sulphide and the sulphur dioxide are converted into sulphur over a bauxite catalyst at a temperature of about 290°C:

$$2\,H_2S + SO_2 \longrightarrow 3\,S + 2\,H_2O$$

The sulphur vapour is then cooled to about 205°C in a waste heat boiler and stored at about 150–170°C.

On start-up, the fuel-gas/air ratio is adjusted according to the required excess of oxygen in the flue gas. The converter bed temperature is then raised as quickly as possible to about 230°C.

The acid gas is fed to the furnace and the combustion-air flow-rate is adjusted so that the stoichiometric ratio of $H_2S:SO_2 = 2:1$ in the tail gas is maintained. The analyser range for $H_2S:SO_2$ ratio is typically 1:1 to 4:1.

In a typical 2-stage sulphur-recovery plant (Fig. 14.16), conversion of the hydrogen sulphide into sulphur is as follows:

Furnace	43%
1st stage converter	35%
2nd stage converter	9%
	___
Total	87%

Thus a range of 0–10 mole % hydrogen sulphide for the tail gas analyser should be quite sufficient.

Two types of analyser are commonly used. Several manufacturers make a special gas chromatograph for this application. DuPont make a special 'colour' analyser system.

As with many analyser systems, the success of the application depends on the design of the sampling system. Here, sampling is very difficult if a true

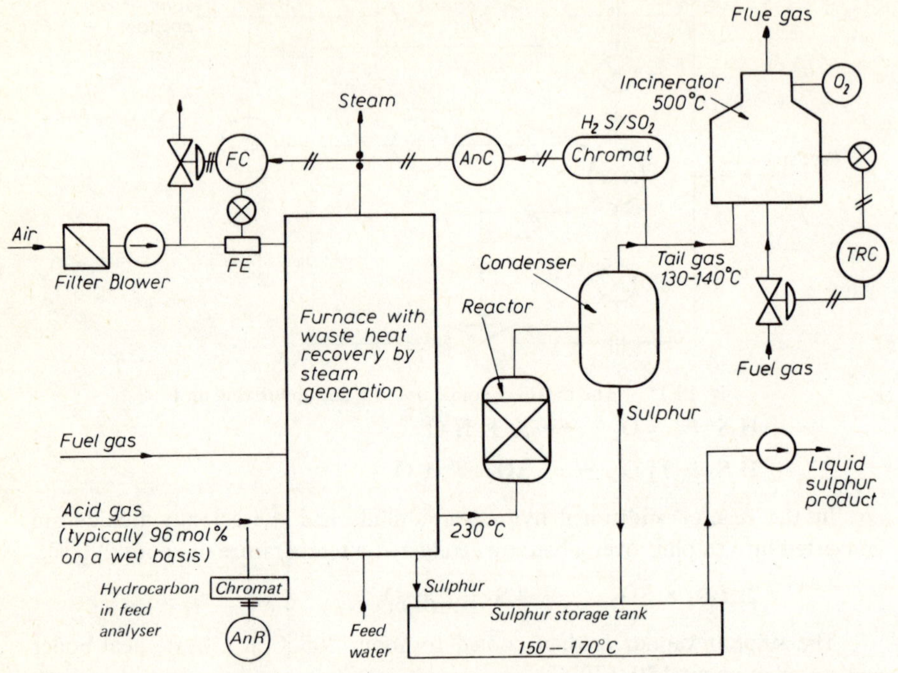

Fig. 14.16 — A two-stage sulphur-recovery plant.

sample is to be obtained. The tail gas is a hot acid containing a high concentration of water vapour. Sulphur condensation also causes problems. The solution is to mount the sampling system and the analyser close to the line. Three hours every week for maintenance on this difficult application is reasonable.

The increased yield of sulphur by control of the $H_2S:SO_2$ ratio in a 200–300 ton/day plant was said to give an analyser pay-off in about half a year. Nowadays the pay-off is due not only to recovered sulphur but also to reduced contamination of the atmosphere and thus to reduced fines paid to the local environment-control authority. It is thus common practice to reduce the concentration of the residual sulphur compounds in the gas feed to the incinerator to about 300 ml/m³ in order to comply with local air-pollution regulations (Fig. 14.17).

In a typical system, the sulphur dioxide in the tail gas from the condenser is reduced to hydrogen sulphide over a catalyst at 300°C. The reduced gas is then cooled and the 'sour' condensate is removed in a quench column. The cooled gas, which typically contains about 3% of hydrogen sulphide and 8% of carbon dioxide is washed counter-currently with a selective amine solution which absorbs most of the hydrogen sulphide but little of the carbon dioxide. The treated gas, which typically contains hydrogen sulphide at about the 300 ml/m³ level, is then passed through a pre-heater before it is oxidized in a cata-

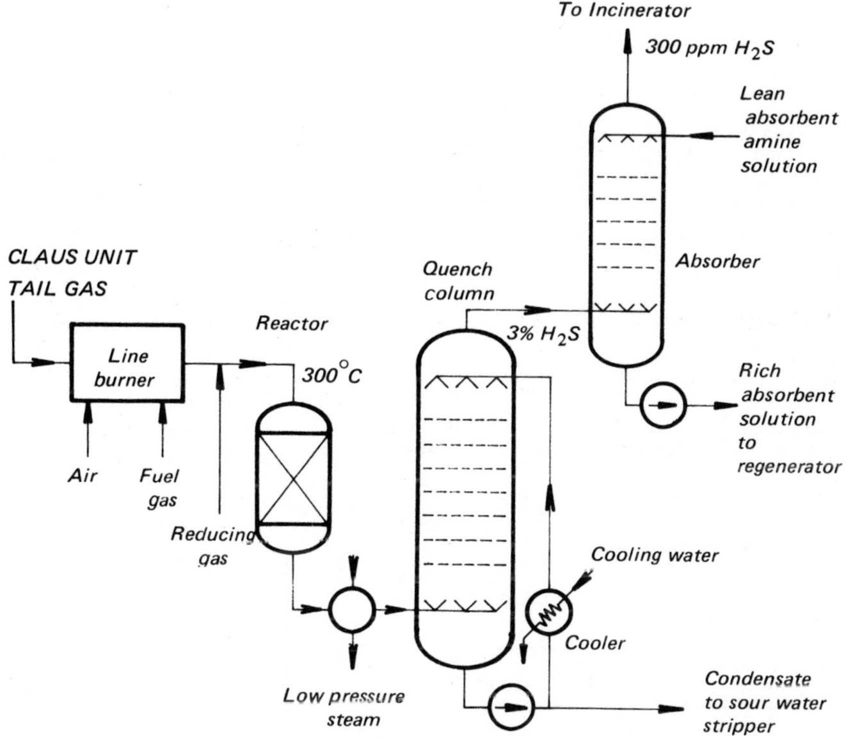

Fig. 14.17 – Tail gas treatment system.

lytic incinerator. The concentration of hydrogen sulphide is then typically less 10 ml/m³.

To check and control the operation of the gas-treatment plant, hydrogen sulphide analysers may be installed to measure the concentration of the hydrogen sulphide in the gas upstream and downstream of the amine absorber and in the incinerator flue gas.

Personnel -protection monitors are usually installed in several places in the plant, especially near to those pumps and valves which, on seal failure, could emit a hazardous concentration of hydrogen sulphide.

14.11 METHANE REFORMING

When relatively pure hydrogen is made by the reforming of methane (Figs. 14.18 and 14.19), each of the reactions is controlled continuously from an analyser signal. Because the analysis is usually required to be done frequently and rapidly, infrared analysers are usually preferred.

However, the economically-priced reliable mass spectrometers that are now available could be used for these applications.

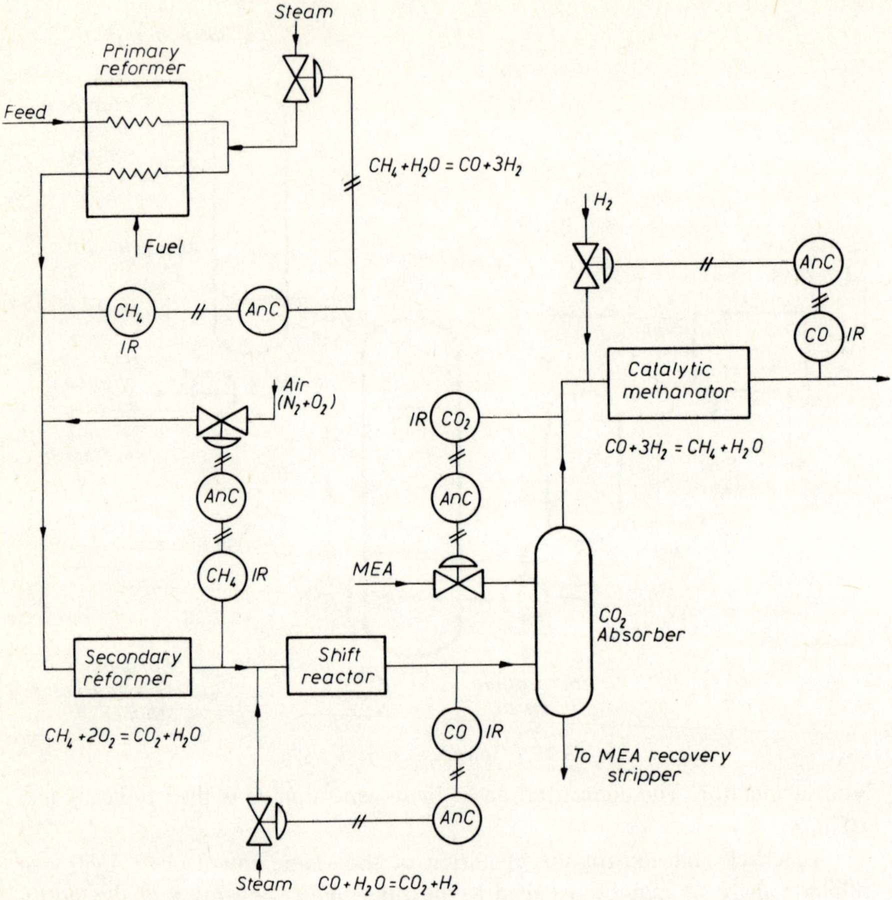

Fig. 14.18 – Methane reformer (MEA = monoethanolamine).

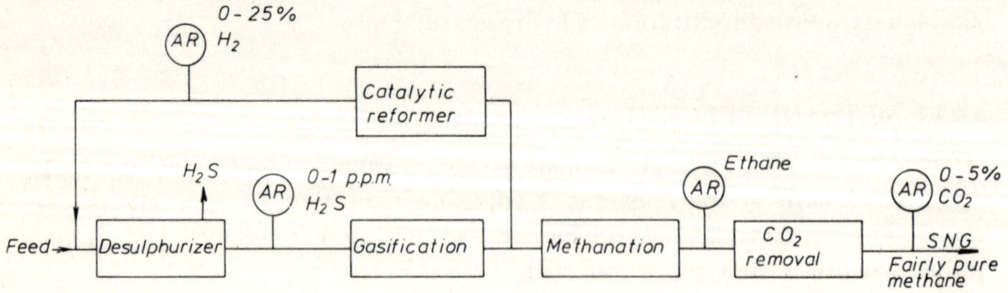

Fig. 14.19 – SNG (methane) plant.

14.12 CARBON MONOXIDE MEASUREMENT BY CONVERSION INTO CARBON DIOXIDE

One way of obtaining a very sensitive measurement of carbon monoxide is to remove interfering gases and oxidize the carbon monoxide to carbon dioxide.

Any carbon dioxide already in the sample stream is first removed by means of a selective chemical scrubber such as sodium hydroxide. The carbon monoxide may then be converted into carbon dioxide with iodine pentoxide in a furnace. The iodine evolved is removed by a chemical scrubber. In the Hilger conversion unit, 50 g of iodine pentoxide will last for about 8 months, for a sample stream flowing at 18 l./hr and containing carbon monoxide at the 100 ml/m^3 level.

14.13 TOP-GAS ANALYSIS IN IRON AND STEEL PRODUCTION

In order for the operation of blast furnaces to be monitored and controlled satisfactorily, analyses for hydrogen, water, carbon monoxide, carbon dioxide, oxygen and methane are required.

Hydrogen in the range 0–5 or 0–10% v/v is determined with a thermal-conductivity analyser. Carbon dioxide is determined to $\pm0.05\%$ v/v in the range 10–15% with a high-precision infrared analyser. The carbon dioxide signal is used to correct the thermal-conductivity signal, which is affected by the carbon dioxide concentration. Carbon monoxide is determined to $\pm0.05\%$ v/v in the range 10–40% by another infrared analyser, and methane in the range 0–1% v/v is determined by a third. Oxygen is determined by a paramagnetic analyser. Nitrogen, if required, is normally determined by difference. Methane is usually determined only when oil-firing is used, and a high-alarm is set to operate when the concentration exceeds a preselected value in the range 0.5–10% v/v.

Mass spectrometers have been used recently to measure the concentrations of all the gases, but there are some problems in their use. Magnetic sector instruments are preferred to quadrupole instruments, since they are more stable and less disturbed by impurities.

Sampling-probe location is critical, since component concentrations vary widely across the stack and according to the distance from the charge. Particulate matter is removed by a hot, dry filter. The clean sample is cooled to a predetermined temperature, so that the remaining gas will have a low moisture content.

The analysis of the waste gases from steelmaking is similar to the analysis of blast-furnace gases, and the sampling procedures are also similar. The most important analyses are those for carbon monoxide and carbon dioxide; continuous monitoring of these concentrations allows the rate of removal of carbon from the bath to be followed and adjusted, if necessary, by adjusting the height of the oxygen lance.

A warning of a dangerous (explosive) concentration of carbon monoxide may be given, allowing the oxygen supply to be turned off as a preventive measure.

The BOS furnace operates with a high $[CO]/[CO_2]$ ratio and the oxygen level is very critical. During the blow with oxygen, the trend in $[CO]/[CO_2]$ ratio is observed in order to control the rate of removal of carbon.

14.14 NUCLEAR REACTOR GAS ANALYSIS

Impurities can build up in carbon dioxide when it is used as a heat-exchange fluid for nuclear reactors. Methane and other hydrocarbon gases will be formed when lube oil leaks into the system and is cracked at the high temperature of operation inside the reactor. The pressure of the boiler-water/steam is much greater than that of the carbon dioxide, so water vapour may leak into the gas. An electrolytic hygrometer may be used to monitor the concentration of the water vapour, which should be about 1 g/m^3.

A gas chromatograph is used to monitor hydrogen (typically 50 ml/m^3), methane (typically 4 ml/m^3) and carbon monoxide (typically 1.5% v/v maximum) if graphite moderators are used. The maximum concentrations of the impurities are limited by blow-down every shift. A sampling frequency of once every 10–15 min is considered adequate.

By monitoring the methane:hydrogen ratio (Fig. 14.20), it is possible to determine the relative importance of the oil and water leaks.

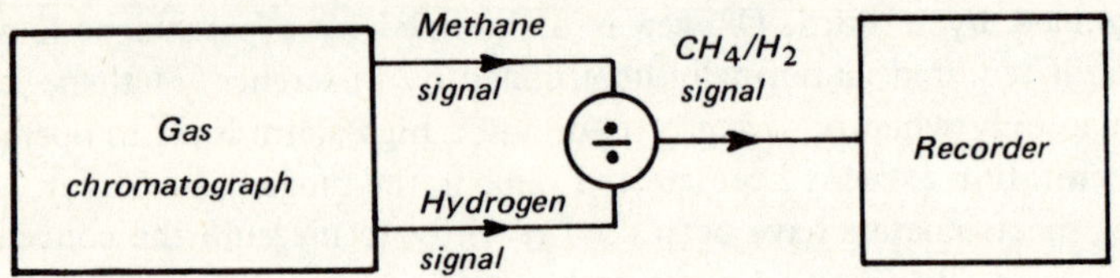

Fig. 14.20 – Methane:hydrogen ratio monitor.

14.15 GASOLINE BLENDING

In a gasoline blender (Fig. 14.21), it is necessary to control the Reid vapour pressure (RVP) of the product by regulating the addition of the higher-vapour-pressure LPG components. It is also necessary to control the octane rating of the product. This may be accomplished first by regulating the amount of re-formate added to the straight run gasoline and secondly by regulating the addition of the lead-alkyl octane-rating improvers (TEL and TML–tetraethyl and tetramethyl lead respectively).

Reformates are being used increasingly for octane-rating improvement instead of the toxic lead alkyls. As a result, the blending is required to be much more precise in order to minimize use of expensive feed components.

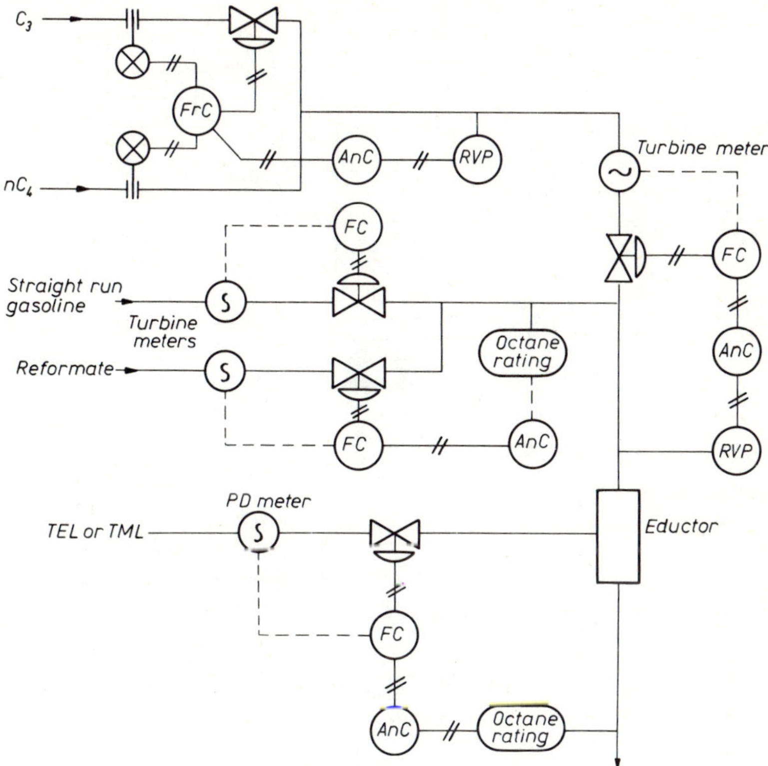

Fig. 14.21 — A gasoline blending system.

14.16 INTERNAL-COMBUSTION-ENGINE EXHAUST ANALYSIS

Sunlight causes the hydrocarbons in the atmosphere to react with oxides of nitrogen to form photochemical smog. A major contributor of hydrocarbons, and other undesirable gases, is the internal combustion engine, so engines must be tested to ensure that they do not emit more than the maximum allowable concentrations of the various pollutants.

The motor vehicle is tested on a roller brake with a programmed cycle of idling, acceleration, running, gear changing and deceleration.

In the 1968 Californian air-pollution regulations, the exhaust gas is required to be analysed instantaneously. The water is condensed out and the dust is filtered out upstream of each analyser.

Two ranges are required for the unburnt hydrocarbons, which are measured by infrared analysers. The low range is 0–1000 mg/kg of hydrocarbons expressed as n-hexane. The high range is 0–10000 mg/kg of hydrocarbons expressed as n-hexane, and this is used during the deceleration part of the cycle when high hydrocarbon peaks occur.

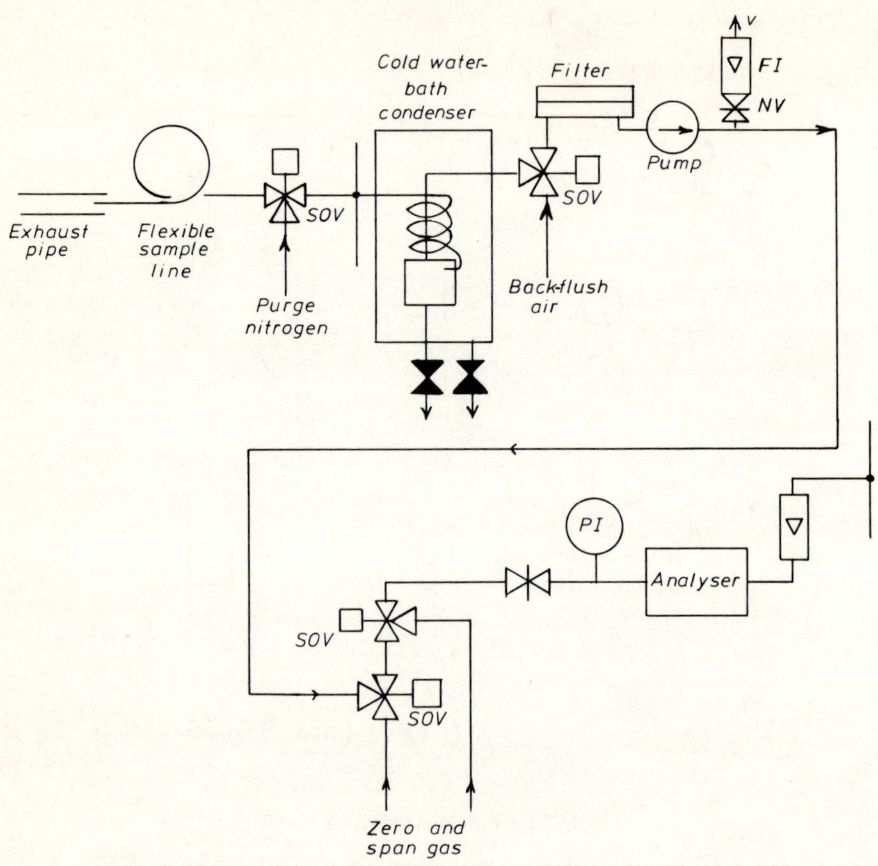

Fig. 14.22 – The California test system.

Cyclonic water separators are supplied with the Horiba infrared analyser, which is modified by Olson for exhaust-gas analysis.

More recently, the average component concentrations have been measured on a weight basis. The exhaust gas is diluted with air to prevent condensation of the large amount of water produced by the combustion of the fuel, and is collected in a bag made from a suitable plastic which has to be selected so that the losses into the bag surface and diffusion through the bag are minimized. Polythene is often preferred.

The concentration of unburnt hydrocarbons is measured in a total-hydro-carbon analyser which uses a flame-ionization detector and has a range of 0–100 mg/kg.

Several types of system are available from various manufacturers such as Beckman, and Hartmann & Braun.

(1) California test system
(2) Europa test (EEC test) system
(3) Constant-volume-sampling test (CVS test) system
(4) SAE diesel-exhaust test system.

In the California test system, there are four non-dispersive infrared analysers with the following ranges:

0–10% of CO, 0–16% of CO_2, 0–1000 mg/kg of HC and 0–10000 mg/kg of HC

The sample is fed to each analyser from the exhaust pipe through a cold water bath condenser, a filter and a pump (Fig. 14.22).

Hartmann & Braun include an infrared analyser with a nitric oxide range of 0–5000 mg/kg and an ultraviolet analyser with a nitrogen dioxide range of 0–500 mg/kg.

Chemiluminescence analysers are also used to measure the concentration of nitric oxide.

In the CVS-test sytem, the exhaust gas is diluted with air (Fig. 14.23). A sample is taken, filtered and passed into a plastic bag. When the bag is full it is disconnected from the sampling system and connected to the analysis system.

Infrared analysers are used for measurement of carbon monoxide and carbon dioxide; a flame ionization detector is used to measure the concentration of unburnt hydrocarbons, typically 0–100 mg/kg (as propane).

In the Federal Register, the US Environmental Protection Agency describes the complete test method.

During a simulated 7.5-mile average trip in an urban area in a 1973 or 1974 vehicle, the maximum allowable average emission per vehicle is 3.4 g of hydrocarbons, 39 g of carbon monoxide and 3 g of nitrogen oxides.

United Nations Regulation 15 describes a method that involves sample bags. For analysis of the unburnt hydrocarbons, an infrared analyser, sensitive to n-hexane, is used.

For all exhaust-gas analysers, it is essential that the reference fuel is specified in detail.

Many manufacturers make simple carbon monoxide and hydrocarbon infrared analysers to improve the setting-up of carburettors.

The Honeywell carbon monoxide measurement/combustion efficiency gradient engine-exhaust analyser (Fig. 14.24) uses a detector which measures the ultraviolet light emitted when carbon monoxide is burned on a hot wire.

14.17 NITROGEN OXIDES

In the absence of sunlight, colourless relatively non-toxic nitric oxide is oxidized to brown toxic nitrogen dioxide. In air, nitrogen dioxide is photodissociated by sunlight into nitric oxide and oxygen. The rate of dissociation increases with temperature to completion at about 600°C.

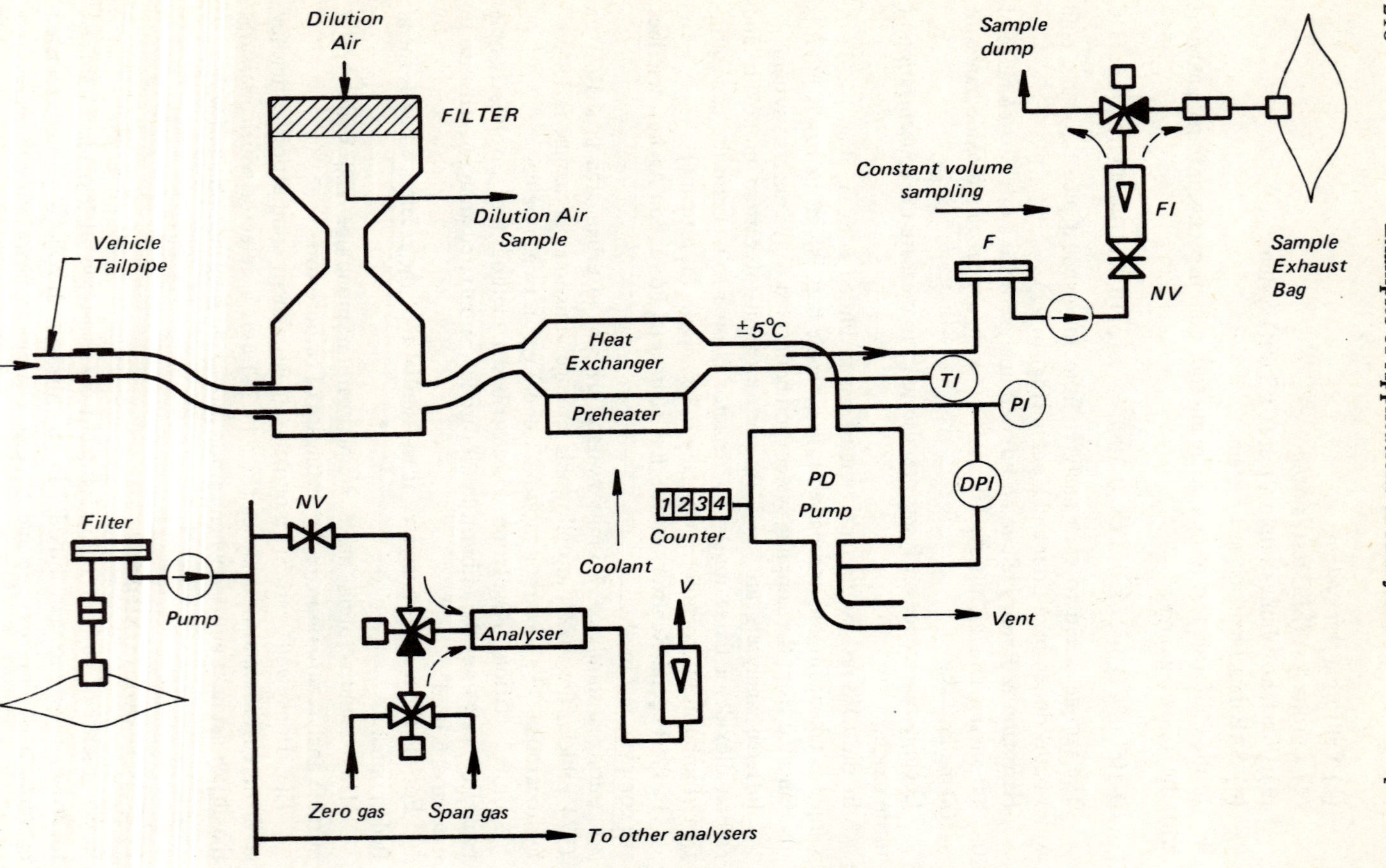

Fig. 14.23 – The CVS exhaust-gas test system.

In order to measure the total concentration of nitrogen oxides, it is necessary either to oxidize nitric oxide to nitrogen dioxide and to measure the concentration of the latter or to decompose the nitrogen dioxide to nitric oxide thermally and to measure the concentration of the nitric oxide.

A photometer working at a wavelength of about 400 nm may be used to measure nitrogen dioxide concentration. To measure nitric oxide concentration, an infrared analyser or a chemiluminescent analyser may be used.

Nitric oxide is readily converted into nitrogen dioxide by aerial oxygen. It is thus necessary to take a sample as near as possible to the process unit if the nitric oxide concentration is required.

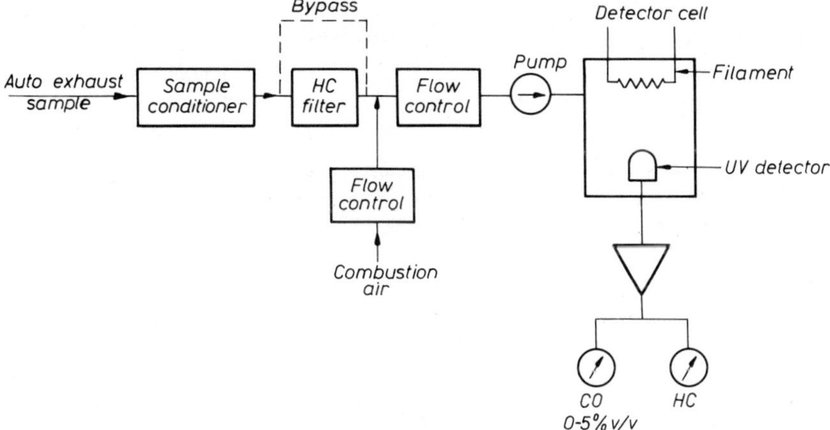

Fig. 14.24 – Honeywell carbon monoxide measurement and combustion efficiency gradient engine exhaust analyser.

14.18 SOUR GAS DENSITY MEASUREMENT

At one refinery, for many years the sour gas density analysers used were found to be rapidly corroded because the gas was too wet. Driers were then added but they did not work satisfactorily.

The total water concentration in the analyser sample gas was 31.8 g/m^3 at 293 K and 1.01325 bar abs, of which 8.1 g/m^3 was the soluble-water concentration, so the free water concentration was 23.7 g/m^3 at this temperature and pressure. Free water, and thus corrosion, was found to occur only when the analyser sample temperature was reduced below 26.7°C (Fig. 14.25).

It was then proposed that the sample be taken from upstream of the pressure-control valve. The sample was cooled, free water was removed and the pressure reduced before the sample was passed to the analyser, which worked at the original gas temperature because it was thermally well connected to the process line (Fig. 14.26). The temperature of the gas was measured so that the gas density under reference conditions could be determined.

The result was that the analyser was available for much longer periods and suffered very little corrosion.

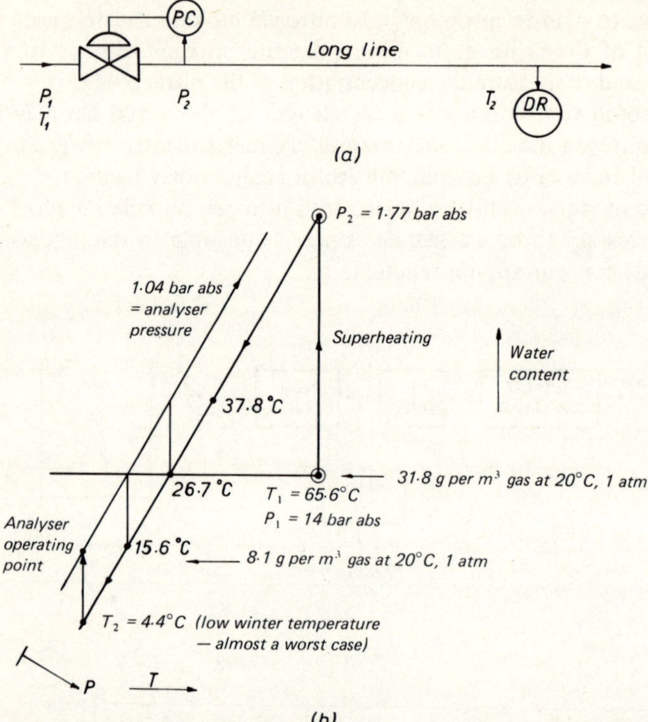

Fig. 14.25 – Sour gas density measurement – incorrect sample-point location.

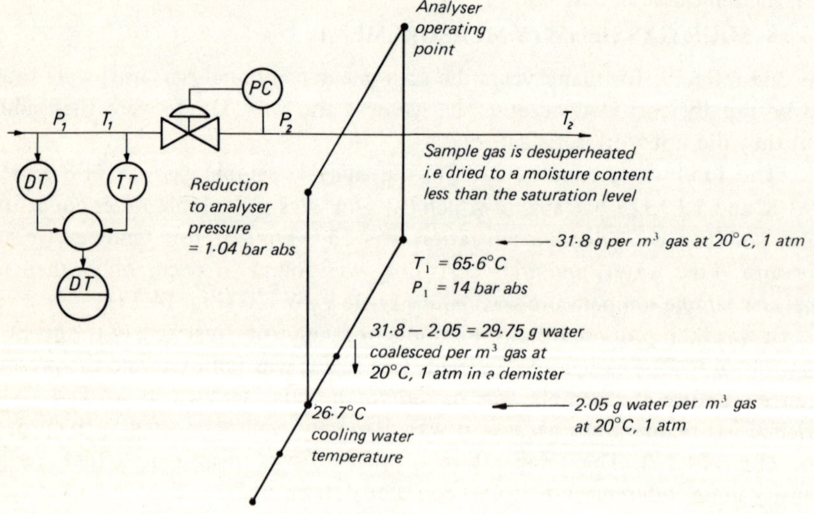

Fig. 14.26 – Sour gas density measurement – improved sample-point location.

14.19 WATER AND SALT IN CRUDE OIL

It is important to know when a well is producing formation water rather than crude oil, in order to minimize the amount of formation water in the crude oil product.

Typically 5–10 wells feed oil into a gas-oil separation production unit (PU) which removes the gas from the oil in three or four stages. The last stage may be a tank or heated stabilizer which finally reduces the vapour pressure of the oil from about 1.5 bar abs to about 0.4 bar abs.

It is not economical to place the interface detector in the well discharge lines; also the vapour pressure is too high. The best location for the detector appears to be in the discharge line of each PU. If the detector finds an unacceptably high concentration of formation water in crude oil, an alarm annunciator should operate. The laboratory analysis records for the previous day for the wells supplying the particular PU should then be examined in order to find which well has started to produce salt.

Two methods are used to measure the concentration of formation water in crude oil.

Salt-in-crude analysers are used for low concentrations, and density analysers are used for high concentrations of water in crude. Nuclear density-analysers could perhaps be used as interface detectors at the wells. Dielectric-constant analysers have been used unsuccessfully for low-concentration measurements.

The formation water contains approximately 20% of salt and has a density of about 1150 kg/m^3. The salt is a mixture of the chlorides of sodium, magnesium and calcium. The product analysers are expected to measure the 10–40 mg/kg range for salt or 40–160 mg/kg for water.

Until 1972, very few salt-in-crude analysers had been sold for this application, and for use in refineries between desalters and crude distillation units for optimization of the former and protection of the latter.

14.20 LUBE OIL AND WAX TREATMENT PLANTS

A simplified block diagram of lube oil and wax treatment plants is shown in Fig. 14.27.

In a refinery lube leg the following plants may be installed:

(i) The viscosity-index improvement plant removes the aromatic oils from the feed. The remaining aromatic content of the essentially paraffinic stream may be measured by a differential refractive-index analyser or a two-wavelength ultraviolet absorption analyser.

(ii) In the oil-dewaxing plant, the feed is cooled to such a low temperature that the wax crystallizes and may be separated from the oil by filtration. Plant operation can be optimized and filter leakage detected by the use of a pour-point analyser.

(iii) The operation of the colour improver can be followed by the use of a colour analyser with a scale according to ASTM D1500.

(iv) Wax de-oiling plants are not common, because of the low demand for refined paraffin waxes. Operation of the plant may be followed by a melting-point analyser and an ultraviolet-absorption analyser.

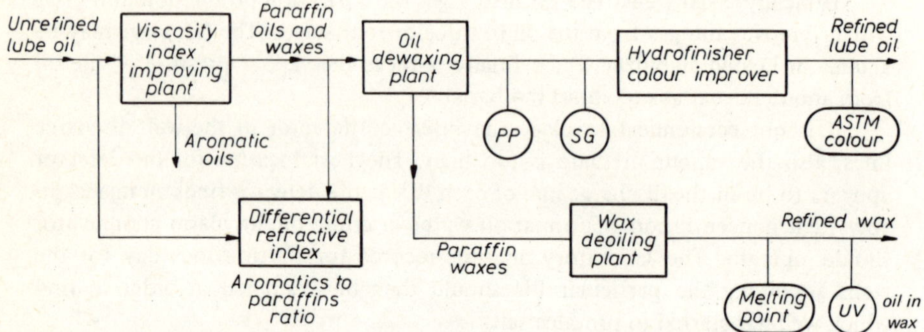

Fig. 14.27 – Block diagram of a lube oil and wax treatment plant.

14.21 POTABLE-WATER CHLORINATION

Chlorination is the most important and widely used method for sterilizing potable water. However, it requires close control to ensure its effectiveness and safety.

The rate of addition of chlorine required for a particular water at any particular time will depend on the following.

(a) The demand for water at that time; this will be very low during the night, but high at peak times during the day.

(b) The pH of the water, and hence the form(s) in which the chlorine dissolves. The reactions are

$$Cl_2 + H_2O = HClO + H^+ + Cl^-$$

$$HClO = ClO^- + H^+ \qquad pK = 7.6$$

Hypochlorous acid is a more effective bactericide than hypochlorite, so a given amount of chlorine will be more effective if the solution is acidic rather than alkaline. The $[HClO]/[ClO^-]$ ratio changes linearly from 9:1 to 1:9 over the pH range 6.6–8.6. At pH $<$ 5 the chlorine remains unhydrolysed.

(c) The chlorine demand of the water, which depends on the bacterial count and on the amount of organic matter present.

It is desirable to ensure that a certain minimum residual level of chlorine remains in the sterilized water (typically 0.2–0.3 μg/ml), first to be sure that the water has been effectively sterilized, and secondly so that bacteria introduced after treatment, e.g. from pipelines, will also be destroyed.

Too much residual chlorine is wasteful and may impart an unpleasant flavour to the water. Ammonia may be added to alleviate this by reacting with the residual chlorine to form relatively tasteless chloramines. Chloramines are less effective bactericides than chlorine, but they are more persistent, and hence offer better prolonged protection against post-contamination of the water.

The ratio between added and residual chlorine concentration depends on the bacterial content or chlorine demand of the feed water and varies from about 4:1 for relatively clear water to about 20:1 for relatively dirty water.

The residual-chlorine analyser allows the chlorination process to be automated. An analyser range of 0–0.5 mg/kg allows the residual chlorine to be controlled to 0.2–0.3 mg/kg. A time-lag has to be built in to the system, because a certain contact time between the chlorine and the water to be treated is required for the sterilization reactions to occur. Sometimes, better control is obtained if the initial dissolved-chlorine concentration is monitored in addition to the residual chlorine: this is particularly true if the water is very dirty.

A typical chlorination system is illustrated in Fig. 14.28.

The residual-chlorine analyser may be adapted for determination of total chlorine by changing some of the reagents.

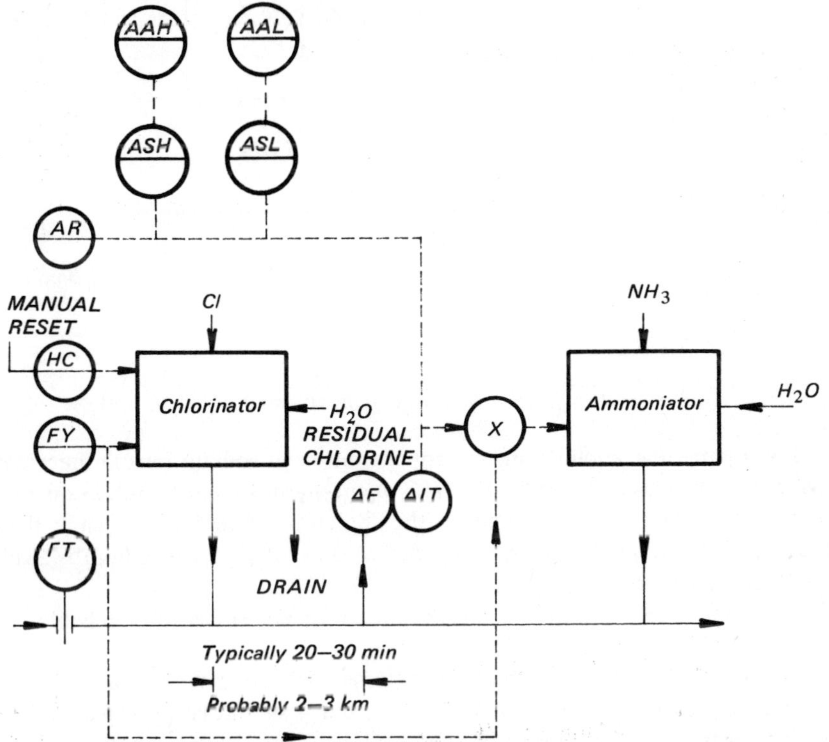

Fig. 14.28 — A typical chlorination system.

14.22 BOILER CONDENSATE MONITORS

The build-up of solids in boilers must be minimized. This is particularly difficult with once-through boilers which have no drum which can be blown through
to reduce build-up.

The usual source of build-up of solids in power station boilers is leakage of
sea-water (or hard water) through condenser tubes. This may be monitored by
the use of a conductivity analyser or by the less reliable but far more sensitive
sodium-ion monitor which uses a sodium-selective electrode (Fig. 14.29).

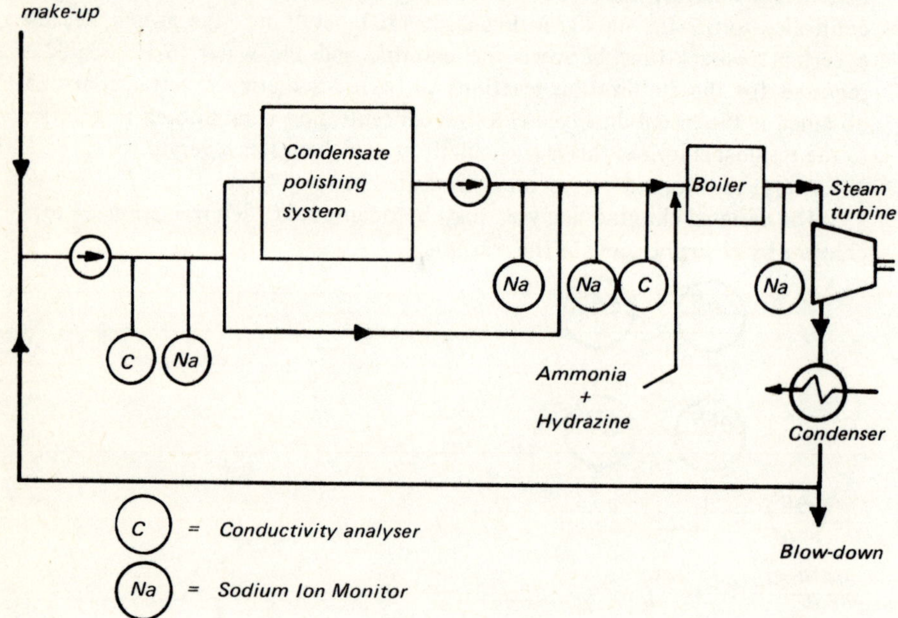

Fig. 14.29 – A dissolved-solids monitor.

For a particular application, the concentration of sodium ions in the water
leaving the condenser should be 1–2 μg/l. A pre-alarm should be set to operate
above 10 μg/l. and the set should be shut down if the leakage is greater than
about 50 μg/l. An instrument with a range of 0.1–100 μg/l. for sodium is available for this purpose.

The sodium-ion monitor is considered to be not sufficiently reliable for
direct connection to the shut-down system.

If it is essential that shut down be automatically controlled by an analyser,
then a conductivity analyser should be used, and it should be provided with an
ion exchanger such as Amberlite IR 120 to make it sufficiently sensitive. It may
be used to measure the concentration of sodium ions or chloride ions, by use of
a different resin. The chloride measurement is particularly important if there is

stainless steel in the boiler tubing. The conductivity ranges used are 0–2 and 0–20 μS/cm.

In order to reduce build-up of dissolved solids, a condensate-polishing system (mixed-bed demineralization system) is included between the condenser and the boiler. This reduces the amount of expensive condensate which has to be blown down.

For optimum reliability, it is advisable to locate the analysers in a clean and quiet dedicated analyser room instead of in a turbine hall.

Reference electrodes are changed every 3 months and sodium-selective electrodes every month. The sodium electrodes can be cleaned and resensitized for reuse.

Calibration is done with sodium-free water obtained from a continuously recirculating mixed-bed ion-exchanger. The accuracy of the standard solution need not be much better than ±5%.

It is advisable for the monitors to be visited once per shift. There is thus no real need for the use of automatic standardization, which is unlikely to give better reliability than a well-trained chemist.

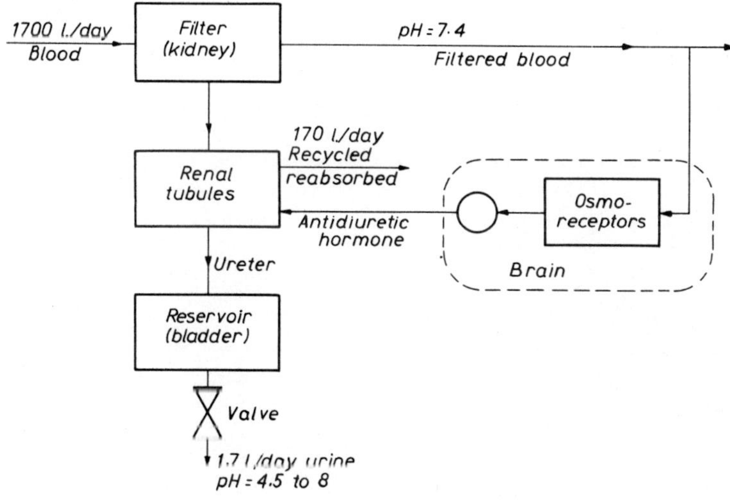

Fig. 14.30 – Control of fluids in the body.

14.23 CONTROL LOOPS IN THE HUMAN BODY

The human body uses several interesting interconnected control loops in which 'analyser' signals regulate various physical and chemical conditions. One such system, which is illustrated in Fig. 14.30, controls the fluid balance in the body.

The following description is based on information received from Dr. R. L. Page.

Osmoreceptors in the brain measure the osmotic pressure of the blood and produce a proportional quantity of antidiuretic hormone which regulates the re-absorption of water in the renal tubules, and the excess of water is discharged as urine.

Any large volume of water (say 1 litre) drunk when the person is thirsty is ingested and dilutes the body fluids, the osmotic pressure of the blood decreases and so the osmoreceptors respond to this by inhibiting the output of the antidiuretic hormone. The re-absorption of water in the renal tubules is diminished and thus the urine output is increased. This increased urine output is maintained until the excretion of the fluid has brought the blood osmolarity back to normal.

CHAPTER 15

Analytical method selection guide

This chapter consists of three parts. Part 1 is a set of annotated contents lists for Books 2–5. Part 2 gives lists of applications of techniques and of techniques for particular applications, and Part 3 lists methods used for analysis of some important materials produced in a wide range of industries.

Part 1 – Contents of Chapters

Book 2 – Quality Measuring Instruments

1. Fluid-density monitors: density of fluids, correlation with other properties such as calorific value.
2. Consistency and viscosity: molecular weight determination.
3. Surface tension: of liquids for optimum atomization and for surfactants.
4. Cloud-point and pour-point analysers: characterization of oils, especially lube oils.
5. Thermal-conductivity gas analysers: purity of light gases.
6. Paramagnetic oxygen-analysers: oxygen in gas mixtures, e.g. in flue gas for combustion control.
7. Dissolved-oxygen analysers: oxygen in water.
8. Vapour-pressure analysers: for gasoline and LPG.
9. Vapour/liquid ratio analysers: high-compression gasolines.
10. Boiling-point analysers: mixtures of liquids, especially of hydrocarbons, to ensure optimum fractionation.
11. Freezing-point analysers: purity of liquids.
12. Calorimeters and flame-speed monitors: to improve combustion.
13. Flash-point monitors: flammable liquids.
14. Octane-rating analysers: control of production of gasoline.
15. Sorptiometry and porosimetry: surface area and pore volume measurements for catalysts, ceramics and cements.
16. Chemical-type analysers: titrations, selective reactions for impurities in liquids and in gases after absorption in water.

Book 3 – Optical Methods

1. Electromagnetic spectra.
2. Emission spectroscopy: elements.
3. Absorption of radiation.
4. Opto-acoustic spectrometry.
5. Refractometry: purity of liquids, moisture content.
6. Polarimetry: concentration of sugar solutions.
7. Tristimulus colorimetry: definition and control of colour, colour body concentration in otherwise colourless liquids.
8. Ultraviolet analysers: unsaturated hydrocarbons, aromatics.
9. Infrared analysers: carbon monoxide, carbon dioxide, hydrocarbons.
10. Turbidity and nephelometry: clarity of liquids such as water, effluents and beer.
11. Fluorimetry and phosphorimetry: aromatics, oil in water, sulphur dioxide.
12. Chemiluminescence: nitrogen oxides.
13. Phosphorescence: oxygen.
14. Raman spectrometry: remote atmospheric analysis.
15. Flame-emission photometry: alkali metals, sulphur, phosphorus.
16. Atomic-absorption spectrometry: many metals.
17. Photoelectron spectroscopy: solid surfaces.
18. X-Ray diffraction: identification of materials and structure determination.
19. X-Ray fluorescence: elements heavier than calcium.
20. Novel techniques.

Book 4 – Electrical and Magnetic Methods

1. Dielectric-constant analysers: purity of fluids, moisture.
2. Electrical conductivity analysers: purity of water.
3. Amperometric analysers: oxidizing gases.
4. Moisture analysers: water content.
5. Potentiometry: acidity, alkalinity, concentrations of ions in water, e.g. sodium, chloride.
6. Coulometric analysers: water content, sulphur dioxide, chloride, sulphur in air, etc.
7. Streaming current detection: control of colloidal systems.
8. Flame-ionization analysers: dilute hydrocarbons in non-combustible gases, GC-detector.
9. Photo-ionization analysers: dilute hydrocarbons in gases, GC-detector.
10. Argon and helium ionization detectors, including electron-capture detectors: GC-detectors.
11. Thermionic analysers: halogen compounds in air.
12. Mass spectrometry: elements, molecules and fragments of molecules.
13. Nucleonic analysers: density, thickness, sulphur, lead, moisture, and many others.

14. Microwave analysers: water in liquids and solids.
15. Nuclear magnetic resonance: protons and ^{13}C-atoms.
16. Electron spin resonance: free radicals.
17. Solid-state monitors: impurities and combustible gases in air.

Book 5 – Gas Chromatographic Methods

Gas–solid chromatography: light gases such as nitrogen, oxygen, carbon dioxide and methane in gas mixtures.

Gas–liquid chromatography: organic molecules in gas mixtures or in completely vaporized liquid mixtures.

Liquid chromatography: organic molecules, especially the less volatile ones, in liquid mixtures.

Part 2

Physical Qualities

Boiling points	2:10
Calorific value	2:12
Cetane rating	2:14
Clarity of liquids	3:10
Cloud point	2:4
Colour	3:7
Density	2:1
Electrical conductivity	4:2
Flame speed	1:12
Flammability	1:9, 4:17
Flash point	2:13
Freezing point	2:11
Octane rating	2:14
Pore volume	2:12
Pour point	2:4
Radioactivity	1:9, 4:13
Refractive index	3:5
Surface area	2:15
Surface tension	2:3
Thermal conductivity	2:5
Turbidity	3:10
Vapour:liquid ratio	2:9
Vapour pressure	2:8
Viscosity	2:2

Elemental Analyses

Atomic absorption	3:16
Chemical analysers	2:16

Emission spectrometry	3:2
Flame emission	3:15
Ion-selective electrodes	4:15
Mass spectrometry	4:12
Nuclear magnetic resonance	4:15
Nucleonic methods	4:13
Solid-state gas monitors	4:17

Molecular Analyses

Chemical analysers	2:16
Fluorescence	3:12
Gas chromatography	5
Infrared	3:9
Liquid chromatography	5
Mass spectrometry	4:12
Ultraviolet	3:8

Gas Analyses

Carbon dioxide and carbon monoxide

Major	2:5
Traces	5, 3:9

Chlorine

In gases	4:3, 2:16
In liquids	2:16

Hydrogen

Major	2:5
Traces	5

Hydrogen sulphide

Traces	2:16, 3:15, 4:17

Moisture

In gases	4:1, 2:16, 4:13
In liquids	2:16, 4:1, 4:2, 4:13, 4:15
In solids	4:14

Nitrogen oxides

Traces	3:12

Oxygen

In gases	2:6, 4:3, 4:5
In liquids	2:7

Sulphur dioxide

Traces	4:6, 3:11

Sulphur compounds
    In gas mixtures   3:15
    In oil   3:19, 4:13

Part 3

Air	2:16, 4:4, 4:8, 4:9, 4:10, 4:11, 4:17, 5
Alloys	3:2, 3:19, 4:13
Beer	2:1, 3:7, 3:10
Boiler feed water	2:7, 2:16, 4:8, 3:10, 4:2, 4:3, 4:5
Crude oil	2:1, 2:16
Fuel oil	2:2, 2:10, 4:13
Gasoline	2:8, 2:9, 2:10, 2:14, 3:8, 4:1, 4:4, 4:13, 5
Hydrocarbon gases	2:1, 2:12, 2:16, 3:9, 4:12, 4:17, 5
Kerosene	2:1, 2:4, 2:10, 2:11, 2:13, 4:13, 5
Lube oil	2:2, 2:4
Margarine	4:15
Olefins	3:8, 3:9, 4:4, 5
Paper	2:2, 4:13
Potable water	2:16, 3:7, 3:10
Slurries	2:1, 3:2, 3:18, 3:19, 4:13
Sugar	2:2, 3:6
Wax	3:8, 4:15

CHAPTER 16

The economic justification for installing an analyser

If an analyser is to be installed to improve the efficiency of the operation of a process unit, it should first be shown that the installation of the analyser system is going to be economically justified. In order to do this, the increased profit made as a result of installing the system must be balanced against the additional cost of installing and running it. The diagram below sets out the main components to be considered in evaluating the economic viability of an analyser.

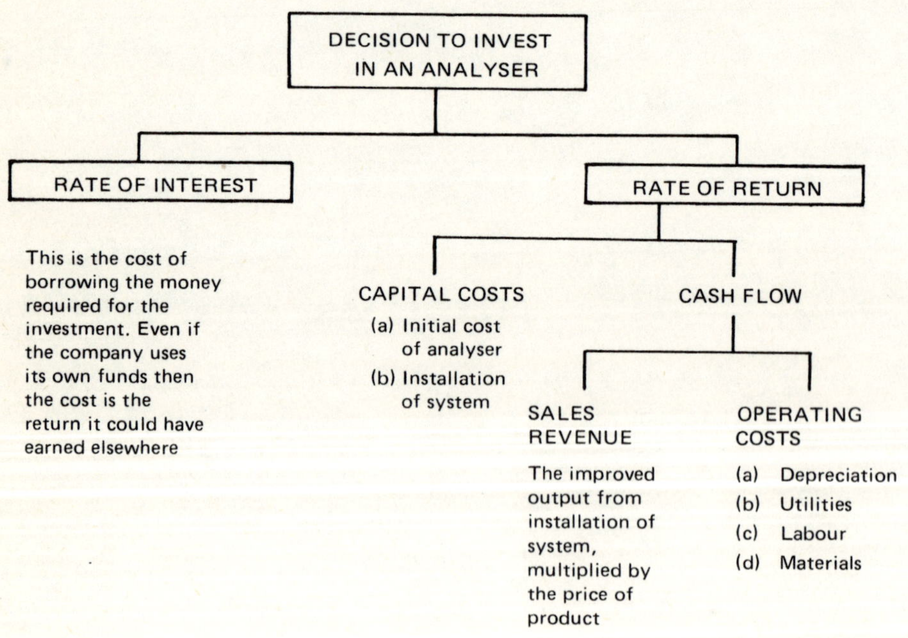

Fig. 16.1

It must be ensured that the rate of return earned on the analyser is at least equal to the cost of borrowing. In order to determine the rate of return on the investment two things must be known: the *capital costs* of installing the system (including the purchase of the analyser and of the remainder of the system that goes with it, along with the installation costs), and secondly, what is called the *cash flow per annum*. In order to calculate the cash flow per annum, an estimate must be made of the effect that the analyser will have on increasing the company's sales revenue. Once a judgement has been made about this, the likely increase in costs involved in installing the analyser should be analysed, for instance, the depreciation of the equipment, the cost of the utilities, and the cost of maintenance (labour and materials). Finally the economic life of the analyser should be estimated. A typical economic life of an analyser is taken to be five years.

Most analysers should pay for themselves within about three years. This pay-back period varies considerably, however, according to the size and type of the process unit. It has been known for analyser systems to pay for themselves in less than a year.

Table 16.1 sets out typical cash flows for an analyser. In this model a number of assumptions have been made.

(1) The economic life of the analyser is five years.

(2) The capital costs involved in buying and installing the analysers are

       (a) initial capital cost = £10,000
       (b) remainder of the system = £3,000
       (c) installation cost = £7,000   Total cost = £20,000

(3) Depreciation is calculated on the basis of the formula:

$$\text{Depreciation} = \frac{\text{Initial cost} - \text{Resale value}}{\text{Life of asset}}$$

If the resale value is £2,000, then the annual depreciation is

$$\frac{£10,000 - £2,000}{5} = £1,600$$

(4) The cost of capital (i.e. interest charges) is assumed to be 20%.

(5) The effects of corporation taxation are ignored.

Table 16.1

A model of typical cash flows for an analyser

| Year | Capital cost £ | Increased sales revenue £ | Operating costs | | | | Net cash flow £ |
			Depreciation £	Utilities £	Labour £	Materials £	
0	−20,000	–	–	–	–	–	−20,000
1		12,200	1,600	800	2,000	300	+ 7,500
2		13.400	1,600	850	1,800	400	+ 8,750
3		14,500	1,600	900	2,000	500	+ 9,500
4		15,350	1,600	950	2,200	600	+10,000
5		15,850	1,600	1,000	2,350	800	+10,100

From the model the following results are obtained:

(1) the pay-back period is 2.4 years,
(2) The discounted cash flow (DCF) rate of return is 33.6%,
(3) Discounting the net cash flows at 20%, the net present value is £6,705.

These three investment-appraisal methods, applied to the data on Table 16.1, all indicate that this particular project could be justified on economic grounds since the rate of return earned on the project is 33.6% and the pay-back period is within the period that is normally considered acceptable for projects of this type.

The cost of engineering, design and project management, which could be of the order of £5,000, has not been included.

It has been assumed that the analyser was installed in an existing analyser house. If it were necessary to share the cost of the analyser house, the cost of the remainder of the system would have risen to about £15,000.

Plant throughput increases as a result of installing the analyser system. Throughput continues to rise as the overall system is better understood. After this initial learning period, increases in plant profit will be due to increases in product values.

The cost of the utilities usually increases by a small but appreciable percentage every year.

The cost of maintenance labour will initially be high until the system is properly understood, when the cost will reduce. It will then increase at a rate dependent on the general increase in labour costs, and as the analyser needs more attention.

The cost of maintenance materials first decreases after teething troubles have been overcome, then begins to rise as the system starts to wear out and as spare parts become more expensive and difficult to obtain. In this analysis, the cost of spare parts purchased but not used has been neglected.

The price of the equipment will probably rise every year. This should not affect the depreciation rate but it will affect the decision for the purchase of a replacement analyser. The decision is usually dependent upon the availability and reliability of the analyser and the resultant maintenance costs. The life of the analyser is normally much shorter than that of the sample system, so each should be considered separately.

Analyser systems installed for reasons other than economic reasons need not be justified in this way. For instance, plant and personnel-safety analyser systems *must* be installed when they are required by the relevant codes of practice.

Acknowledgement

The author acknowledges the help of W. B. Hornby, Senior Lecturer in Economics at The School of Business Management Studies, Robert Gordon's Institute of Technology, Aberdeen, Scotland.

Bibliography

Davies, J. R. and Hughes, S., *Managerial Economics,* Macdonald and Evans, London, 1977.

Merrett, A. J. and Sykes, A., *The Finance and Analysis of Company Projects,* Longmans, London, 1963.

Hawkins, C. J. and Pearce, D. W., *Capital Investment Appraisal,* Macmillan, London, 1971.

Index